Dangerous Snakes of Australia

A Guide to Their Identification, Ecology, and Conservation

Mike Swan

Hellbender Publishing
A Zona Tropical Publication

Comstock Publishing Associates
an imprint of
Cornell University Press
Ithaca and London

Photographs used by permission of the respective photographers.

First published 2024 by Cornell University Press

Printed in China

Librarians: A CIP catalog record for this book is available from the Library of Congress.

ISBN 978-1-5017-7549-9 (paperback)

Book design: Gabriela Wattson

Front cover:
  Mulga Snake (*Pseudechis australis*), Ken Griffiths

Back cover (left to right):
  Yellow-bellied Sea Snake (*Hydrophis platura platura*), H.G. Cogger
  Desert Banded Snake (*Simoselaps anomalus*), B. Schembri
  Highlands Copperhead (*Austrelaps ramsayi*), G. Wallis
  Black-striped Burrowing Snake (*Neelaps calonotos*), B. Maryan
  Dunmall's Snake (*Glyphodon dunmalli*), G. Stephenson

Contents

Preface

In 1952 Australian herpetologist Eric Worrell produced the first handbook on the dangerous snakes of Australia. It was designed as a guide for use by people who lived in the bush, bushwalkers, missionaries, servicemen and -women, Boy Scouts, Girl Guides, migrants, and naturalists.

It was a very different Australia in those days, with many people living in less-developed areas and encountering dangerous snakes on a regular basis.

Worrell's handbook provided information on the known dangerous species, which at the time numbered about twelve land snakes and a few sea snakes. It also listed a procedure for first aid after a snakebite. Surprisingly, at the time, only about five persons per year died from snakebite, despite the dozens of recorded bites. First aid treatment was very basic and harsh, with the minimal recommended equipment being a "new" razor blade and a ligature. Those were also very early days in the production of snakebite antivenoms, which had begun experimentally in the last decade of the nineteenth century.

Our knowledge of Australia's dangerous snakes and the treatment for snakebite have come a long way, and the described number of species has increased considerably.

This book has been produced to allow readers to identify all dangerous Australian snakes using descriptive species accounts, photographs, and distribution maps. The species accounts provide detailed morphological features of each dangerously venomous, medically significant, and potentially dangerous species, with notes on lethality, habitat, behavior, and distribution. The IUCN (International Union for Conservation of Nature) conservation status is listed for each species. For dangerously venomous species, there is a section covering snakebite, snake venoms, and antivenom.

In concert with the continuing loss of worldwide biodiversity, Australian snakes are at risk from changes to habitat through overgrazing, clearing, and controlled burning, and from invasive species preying on their eggs and juveniles, resulting in low recruitment (addition of new individuals to a population).

Hopefully, through this publication, a much better understanding of the distinctive habits and lifestyles of the dangerous snakes of Australia and their low risk to the safety of human beings will lead to more concerted efforts for their long-term conservation.

Acknowledgments

The subject of this book is one I am very familiar with, having spent much of my life around dangerously venomous snakes, and I feel a great sense of satisfaction upon completing a book about this misunderstood group of reptiles. Becoming an author and having your work accepted for publication is not an easy process, but it does become easier when friends, colleagues, and family provide their support.

I would like to thank Scott Eipper, John McCuen, Greg Parker, Chris Williams, and Steve K. Wilson for their initial support for this project.

The images used throughout this book were supplied by many of Australia's leading herpetological photographers. Gary Stephenson and Steve K. Wilson were particularly generous with their images, and I thank them for that.

The following people also kindly provided their images: Luke Allen, Shane Black, Jack Breedon, Adam Brice, Brian Bush, Natalie Callanan (Pilbara Dive and Tours), Jesse Campbell, Michael Cermak, Angus Cleary, Hal Cogger, Scott and Tie Eipper (Nature 4 You), Jules Farquhar, Nick Gale, Prathamesh Ghadekar, Ken Griffiths, Alex Holmes, Paul Horner, Max Jackson, Stephen Mahony, Connor Margetts, Brad Maryan, Jake Meney, Greg Parker, Arne Rasmussen, Akash Samuel, Brendan Schembri, Shawn Scott, Ollie Sherlock, Raymond Sillett, Ruchira Somaweera, Peter Street, Serin Subaraj, Jason Sulda, Jordan Vos, Greg Wallis, David Williams, Justin Wright, and Anders Zimny.

For assistance with inquiries regarding the implementation of lethality ratings I am grateful to Tim Jackson, Peter Mirtschin, Glenn Shea, Rick Shine, and Christina N. Zdenek.

Rachael Hammond provided the identification illustrations, and Marcus Whitby assisted with distribution maps.

Amy K. Hughes performed an intensive and highly professional edit of the book.

Bernie O'Keefe has been great company on numerous recent field trips.

My wife, Stephanie, and sons, Daniel and Timothy, have always provided encouragement for my herpetological endeavors, and I thank them for their immeasurable support.

Introduction

Australia has the reputation of being a land filled with dangerous creatures, which include crocodiles, sharks, spiders, and some of the most dangerous snakes in the world. It is true that there are numerous species of dangerous snakes in Australia, but there are also other types of snakes that pose little risk to human beings. Pythons are the largest living snakes, and Australia has fifteen species and six poorly defined subspecies. They have solid teeth, lack any venom apparatus, and utilize constriction to subdue their prey. Even the largest Australian python poses little risk to human beings. Colubridae, the largest family of snakes in the world, has only a handful of species in Australia. These recent invaders from Asia include the Keelback (*Tropidonophis mairii*), the Slaty-gray Snake (*Stegonotus australis*), and a couple of diurnal tree snakes. These species all lack fangs and are harmless. Another colubrid, the Brown Tree Snake (*Boiga irregularis*), is a nocturnal, venomous, mostly arboreal snake with fangs located in the back of the jaw instead of up front. It bites readily if disturbed, and a bite can cause some local discomfort, but it is not considered dangerous.

Two species of file snakes (family Acrochordidae), so named because of their rough skin texture, occur in billabongs and brackish environments of tropical Australia. These nonvenomous snakes with loose baggy skin and blunt snouts have served as an important food source for indigenous people. Mangrove snakes (family Homalopsidae), also known as mud snakes, are rear-fanged snakes that inhabit mangroves and freshwater streams. They are distributed from the Gulf of Oman through Asia, New Guinea, and parts of Micronesia, and five species occur in tropical northern Australia. All species are venomous, though bites from these snakes are not considered dangerous. Another group, this one poorly understood and occurring throughout Australia, is the blind snakes (family Typhlopidae). These are completely harmless burrowing snakes that are observed mostly at night when they come to the surface after rainstorms. Australia has forty-eight described native species and one introduced Asian species.

Australia's dangerous snakes are front-fanged—with fangs located at the front of the mouth—and are part of a large group of advanced snakes related to the well-known cobras and mambas of Asia and Africa. All belong to the family Elapidae, the most dominant Australian snake family, which also includes sea snakes. Most species have a complex venom production and delivery system. In Australia, elapids range in size from small burrowing species like Jan's Banded Snake (*Simoselaps bertholdi*), with a maximum length of 300 mm, to the large, formidable Coastal Taipan (*Oxyuranus scutellatus scutellatus*), which may reach 3 m.

Lethality

In Australia the family Elapidae (front-fanged venomous snakes) comprises 109 species, two represented by two subspecies each, of terrestrial snakes and 32 species of marine snakes. They are distributed through a variety of habitats across mainland Australia, Tasmania, and surrounding waters.

The terminology used to describe the lethality of these species is very inconsistent and requires standardization. Complicating things further, the venom potency of many less-significant species has not been studied in detail; however, current research has shown that some genera are represented in more than one lethality category.

The categories implemented in this book are:

Dangerously venomous: capable of delivering a bite to a human that, if untreated, has a high likelihood of a fatal outcome (33 species, one with two subspecies, of land snakes and 24 species of marine snakes)

Medically significant: capable of delivering a bite to a human that, if untreated, is likely to have a serious and possibly fatal outcome (11 species of land snakes)

Potentially dangerous: unlikely to deliver a lethal bite to a human, based on current evidence or by inference from closely related species that have been studied, but a bite may produce clinical signs, potentially very serious, in some individuals (65 species, one with a subspecies, of land snakes)

Australia's Dangerously Venomous Land Snakes: Main Groups

Death Adders (genus *Acanthophis*)

Death adders belong to the genus *Acanthophis* and are widely distributed from coastal forested areas to sand dunes and spinifex habitats of central Australia. They are viper-like in appearance, with short, fat bodies and thin tails. Death adders are ambush feeders that shelter in leaf litter and beneath vegetation. These cryptic snakes are not inclined to move if disturbed and present a danger if trodden on. They have highly toxic venom and large fangs and can strike with rapid bites in quick succession.

Copperheads (genus *Austrelaps*)

The southeastern cooler, forested regions of Australia contain the Highlands Copperhead (*Austrelaps ramsayi*) and Lowlands Copperhead (*A. superbus*). A third species, the Pygmy Copperhead (*A. labialis*), occurs in woodlands and agricultural areas of southeastern South Australia. The Lowlands Copperhead also occurs in Tasmania and some Bass Strait islands. Copperheads are very cold tolerant and thus the last snakes to retreat for the winter and the first to venture out in the spring. They have very toxic venom but are inoffensive snakes and disinclined to bite.

Tiger Snake (*Notechis scutatus*)

The Tiger Snake (*Notechis scutatus*), until recently, comprised a variety of subspecies. Despite these snakes' varied distribution and different forms, genetic work has shown them to consist of a single species. Tiger Snakes are distributed through a variety of habitats from Victoria to southeast Queensland, Tasmania, some Bass Strait islands, South Australia, and southwestern Western Australia. The mainland forms are associated mostly with waterways, including rivers, creeks, and swamps. They shelter beneath ground debris like logs and rocks and are sometimes observed basking close to a retreat. Tiger Snakes are relatively inoffensive snakes.

Taipans (genus *Oxyuranus*)

The Coastal Taipan (*Oxyuranus scutellatus scutellatus*) is found in coastal areas from northern New South Wales to northern Queensland and across the Top End of the Northern Territory to the Kimberley region of Western Australia. Another subspecies occurs on islands in the Torres Strait. Both the Inland Taipan (*O. microlepidotus*) and the Western Desert Taipan (*O. temporalis*) are restricted to inland arid zones. These large species feed exclusively on mammals and are considered among the most lethal snakes in the world.

Black Snakes (genus *Pseudechis*)

This group of moderate to large species is generally referred to as "black snakes," though only a few species are actually black. The Mulga Snake (*Pseudechis australis*) is widely distributed in Australia in varying color forms. It attains a large size (up to 2.5 m) and occurs in a variety of habitats. Though it is a reasonably inoffensive species, the venom glands contain large quantities of venom, and it can deliver a very dangerous bite. A number of other species of "black" snakes, associated with particular habitats including black-soil plains, occur through drier regions of Australia. The Red-bellied Black Snake (*Pseudechis porphyriacus*) occurs mostly in association with wetlands in eastern Australia, and the Papuan Black Snake (*Pseudechis papuanus*) is recorded in Australia only from the Saibai and Boigu Islands, in the far northern Torres Strait.

Brown Snakes (genus *Pseudonaja*)

Brown snakes occur throughout most of Australia and are associated more with arid habitats. While they may shelter beneath ground debris and vegetation, they utilize holes in the ground particularly in areas where little ground cover exists. All are nervous species with toxic venom and if threatened usually raise their forebody in a double S position and may advance toward the threat with mouth open. The Eastern Brown Snake (*Pseudonaja textilis*) is responsible for most snakebites in Australia.

Rough-scaled Snake (*Tropidechis carinatus*)

The Rough-scaled Snake occurs in subtropical coastal environments from central New South Wales to Fraser Island (K'gari) in Queensland and in a separate distribution within the Wet Tropics of northern Queensland. It is associated mostly with rain forest and wetlands and is a highly venomous and extremely dangerous snake.

Dangerous Snakes and People

Australians who live in centers of large cities have little chance of encountering dangerous snakes; however, as one moves outward into the suburbs, contact with dangerous snakes becomes a real concern. Tiger Snakes can be found relatively close to the central business district of Melbourne, Victoria, along the Yarra River and are commonly encountered. In Brisbane, Queensland, people are more likely to have a harmless Coastal Carpet Python (*Morelia spilota variegata*) or Common Tree Snake (*Dendrelaphis punctulatus*) in their garden, but in some areas the very defensive and dangerous Eastern Brown Snake is observed frequently. Sydney, New South Wales, is Australia's largest city and hosts a great variety of snakes throughout the metropolitan area, including dangerously venomous species like the Eastern Brown Snake, Red-bellied Black Snake, and Common Death Adder (*Acanthophis antarcticus*).

People living in houses situated close to bushland, parks, and conservation reserves are more likely to have snakes enter their property seeking food and shelter. It is advisable to keep properties clean and tidy, without building materials lying on the ground. Sheets of iron, timber, roof tiles, and concrete slabs provide good refuge for snakes. Rats and mice are strongly favored prey items, and care should be taken to avoid providing breeding opportunities for these introduced pests.

Work huts and Western Brown Snake (*Pseudonaja mengdeni*), Pilbara, WA

It is a difficult task to keep snakes from entering properties, and a smooth high wall is the only real preventative measure. Snake repellents that emit ultrasonic vibrations into the ground do not work, nor do liquid concoctions based on the scents of various plants. The use of bird netting means snakes become entangled, resulting in an extremely dangerous situation for people and a slow, cruel death for snakes. It is important that doors and windows are sealed properly with screens to ensure snakes do not enter houses. During hot weather it is advisable to carry a torch (flashlight) at night, as some dangerous snakes are nocturnal.

Interactions between dangerous snakes and humans (and also livestock, cats, and dogs) are inevitable. Fortunately, the chances of being bitten by a dangerous snake in Australia are extremely low, and antivenom is available to treat the bites of all species.

Identification of Dangerous Snakes

Unless you are an experienced herpetologist or very familiar with snakes, it is not necessarily a straightforward process to correctly identify different species. However, correct identification is important for the management of snakebite, as it is always desirable to verify the species of snake responsible for the bite.

In addition to the many snake species in Australia, there are forty-six described species of harmless legless lizards, and these are often mistaken for small snakes. The differences between snakes and legless lizards are numerous. Snakes have a deeply bifurcated tongue, no ear openings, a short tail about one-third the length of the body, usually wide ventral scalation, and a flexible jaw structure. Legless lizards have a broad, fleshy tongue, usually external ear openings, a long fragile tail about two-thirds the length of the body, narrow ventral scalation, small scaly hind-limb flaps, and a rigid jaw structure.

Painted Delma (*Delma petersoni*), legless lizard

Peninsula Brown Snake (*Pseudonaja inframacula*), juvenile

Terrestrial elapid snakes have enlarged, symmetrically arranged scales on the top of the head; large, laterally expanded ventral scales; and a cylindrical, pointed tail. They can be separated from the similar-looking colubrid snakes by the lack of a loreal scale between the nasal and preocular scales. Marine elapid snakes are most easily recognized by their flattened, paddle-shaped tail. Scalation (the arrangement, number, and character of the scales) and distribution also must be taken into consideration when identifying snakes.

Head scalation in elapid snake

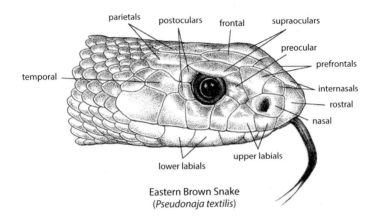

Eastern Brown Snake
(*Pseudonaja textilis*)

Snake mid-body scale rows: diagonal counts

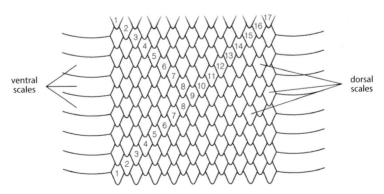

Snake ventral area with single and divide anal and subcaudal scales

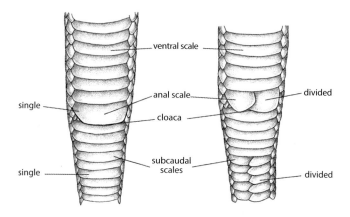

Counting scales is an important method used to identify snakes as scale counts vary considerably between species. Mid-body scale rows on the snake's dorsum (back, or upper surface) are counted diagonally in the mid-body region (the widest part of the body, approximately halfway down the snake's length) from ventral scale to ventral scale.

The number of ventral scales is also important; they are counted from beneath the head, at about the same point as the back of the jaw, to the cloaca. The subcaudal scales are those from the cloaca to the tail tip; they may be single (undivided) or divided. The anal scale, also useful in identification, may be single or divided.

The total length (TL) of a snake is measured from the snout to the tail tip.

A snake should always be considered dangerous until proved otherwise. Dead snakes should be handled with caution and preferably only by the tail. Handling dangerous snakes for identification should be attempted only by experienced people. The safest restraint method is to place the head and a portion of the body into a clear plastic tube that is about the same diameter as the snake.

Snakebite

An estimated 5 million snakebites occur globally every year, resulting in 138,000 deaths and leaving 300,000–500,000 people disabled.

Australia is home to about 26 million people, and 3000 bites from snakes are reported annually, resulting in 300–500 hospitalizations and two or three deaths. These statistics from the Royal Flying Doctor Service of Australia include people who have been accidentally bitten and also people who put themselves in harm's way by interfering with snakes (sometimes when alcohol-affected).

A small number of reptile keepers and snake removalists are also bitten each year by captive dangerous snakes and by snakes being removed from people's properties; 30–40% of people are bitten around their own homes.

The number of people who die from snakebite each year in Australia is very low, and encounter rates with snakes are also low, in comparison to those in other parts of the world, especially Asia, where snakebite constitutes a serious threat to human life. In India, about 50,000 people die annually from snakebite. On the island of Sri Lanka, about 80,000 people get bitten by snakes each year, of which about 400 lose their lives. In Australia, the use of protective footwear, the availability of high-quality antivenom, and efficient medical services result in a very low number of human deaths.

Asia, Africa, and South America have very dangerous species of vipers (family Viperidae), which are relatively large, defensive, and regularly encountered. Snakebite is common in agricultural lands due to limited preventative measures, including wearing appropriate footwear and proper first aid training. Vipers are absent from Australia.

The chances of being bitten by a dangerous snake in Australia are low, and antivenom is available to treat all species of dangerously venomous snakes. Sudden death from snakebite is unlikely; an average time to death is twelve hours, which in most cases should provide ample time to implement appropriate first aid procedures.

Snakebite management plans, including first aid, should be refreshed each season (as with a flood- or fire-evacuation plan or CPR training).

Snake Venom

Front-fanged venomous snakes are classified as proteroglyphs: they have a pair of fixed immovable fangs located at the front of the mouth, and most species have a complex venom production and delivery system. A large venom gland is situated in the temporal region of the head, connected via a duct running along the upper jaw just beneath the skin to the base of a large, syringe-like fang on the anterior end of the maxillary bone. Elapid snakes have fixed, hollow, tubular fangs formed by a continuation of the dentine across the anterior seam.

The venom gland produces a secretion containing a mixture of toxins and other substances. Some of these toxins can block the chemical communication from the nerves to the muscles in vertebrates, leading to paralysis and death. This specialized buccal-gland secretion may be used to obtain prey and in self-defense. It is also possible that snake venom plays some role in the digestion of prey items.

Coastal Taipan (*Oxyuranus scutellatus*), skull

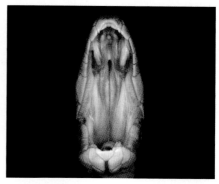
Coastal Taipan, mouth with fang sheaths

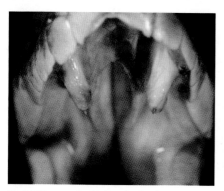
Coastal Taipan, mouth with fang sheaths, close-up

Coastal Taipan, replacement fangs

All known snakes are strictly carnivores; some species consume invertebrates, but most venomous snakes prey upon small vertebrates, primarily reptiles, amphibians, birds, fishes, and mammals. For most species, the size of prey items is restricted to animals that can be swallowed whole, as snakes are not capable of chewing food. The function of snake venom appears to relate most directly to the securing of prey; one strategy is to paralyze muscles and restrict the ability of the prey item to escape.

Australian snake venoms are among the deadliest in the world, containing complex mixtures of biologically active proteins and other substances. Many snake venoms contain neurotoxins that can react rapidly on the nervous systems of small animals, particularly mammals and birds. Neurotoxins also affect the muscles involved with the inflation of the lungs, causing a blockade of the neuromuscular junctions in the diaphragm and resulting in respiratory failure. Other toxins can damage muscles, blood vessels, and body tissue. Some Australian elapid venoms are capable of rapidly producing an anticoagulant effect, depleting the blood-clotting process. Hemolytic venoms destroy red blood cells, resulting in intravascular hemolysis. Cytotoxins destroy cells in the blood and tissue and include cardiotoxins (affecting the heart) and myotoxins (affecting muscles).

Australian snake venoms have been researched extensively, with tiger snake venoms probably the most studied. The toxicity of venom is determined using the LD_{50} (lethal dosage 50) test, which defines the lethal dose (LD) of a substance that will kill half of the test animals. Typically, this method requires 100 or more test animals. The toxin is administered in increasing doses, usually five or more doses per animal, to groups of ten male and ten female mice. The mortality rate within a given period is recorded, and the LD_{50} is determined with the aid of statistical calculations.

Australia's Inland Taipan is an impressive species, and it became infamous when, in 1979, it topped the LD_{50} test and was found to have the most potent venom of any snake in the world (Broad, Sutherland, and Coulter 1979). A lot of other well-known Australian venomous snakes were also tested, along with a selected few species from other regions. This resulted in a list of the most venomous snakes in the world, but it does not necessarily make them the most dangerous snakes. The LD_{50} test is performed on mice, not humans, and it does not give a true representation of how snake venom works on humans.

The development of this list sparked a whole new world of research, centered on the premise that Australia's snakes, due to their high toxicity values, are the most dangerous in the world. What this biased study failed to highlight was that it didn't include many well-known highly dangerous snakes from other continents and, perhaps even more important, the study had little relevance to humans as it was conducted on mice.

The LD_{50} test is purely academic, meant to compare venoms with a standardized test. The results show little relevance to what would happen to a human envenomated by these snakes. Venom has evolved to be most effective on a snake's preferred prey items, so if the venoms were tested on frogs, birds, lizards, or any other prey type, the results would likely favor the snakes that prefer that particular prey. It is no coincidence that mice are the preferred prey of most of the top-ranking snakes on the list.

It is problematic to try to determine Australia's most dangerous snake simply based on toxicity studies. Agonizing over which species are more toxic is somewhat irrelevant, as the dangerously venomous Australian elapid snakes are quite capable of killing a human being, and there are no degrees of "deadness." To interpret the dangerousness of Australian snakes more clearly, scientists developed a more relevant concept based on the actual threat posed to human lives. Brown snakes and tiger snakes are no doubt at the top of the list, as they are relatively common in urban areas and often encountered. There are few reported bites from sea snakes, as they are generally quite timid and inoffensive.

The potential danger posed by juvenile and subadult snakes should not be underestimated, as they may not be able to control the amount of venom released. In 2001, American herpetologist Joseph Slowinski died from the bite of a small Suzhen's Krait (*Bungarus suzhenae*) while researching snakes in Myanmar. There are many other factors that come into play in the release of venom, including species, body size, hydration, diet, and when the snake has last fed—but this is relevant to snakes of all life stages, not just juvenile snakes.

Considering the immense amount of energy and substance it takes to create venom, most snakes can control and decide when to utilize it. Often, they will dry bite as a defense mechanism rather than waste their precious venom.

Antivenom

During the nineteenth century, the reported cases of snakebite in Australia were increasing, so new and dramatic snakebite treatments were explored by self-styled "snake men" using diverse remedies. Many would inflict on themselves multiple bites from highly dangerous snakes to either prove a certain treatment or add more thrills for an audience. The mortality rate among these snake handlers and showmen was high.

It was not until the last decade of the nineteenth century that an experimental approach (led by Thomas Lane Bancroft in Queensland and Charles James Martin in Sydney and Melbourne) brought scientific objectivity to the subject of snakebite and ushered in the modern era of Australian toxinology. This was accompanied by a more detailed evaluation of the behaviors and lifestyles of all the known dangerous species.

Medical research in Australia from 1895 to 1905 coincided with the emergence from Europe and the Americas of some new knowledge of the therapeutic effects of antitoxins. The subsequent systematic study of Australian venoms and toxins, through the 1930s and beyond, by Frank Tidswell, Neil Hamilton Fairley, Ian Clunies Ross, Sir Charles Kellaway, and

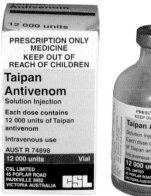
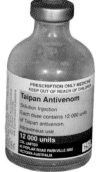

Coastal Taipan, antivenom

John Burton Cleland set the foundation for Australia's leading reputation in venom research. These developments revolutionized the medical management of snakebite victims, who in the past had died from bites by dangerously venomous snakes. Frederick Grantley Morgan, Jack Graydon, Saul Weiner, William Lane, and Harold Baxter at the Commonwealth Serum Laboratories in Melbourne emphasized the importance of cooperation between snake handlers skilled at collecting and milking venomous snakes and those developing the antivenoms. In the development and early production of antivenoms, the collecting of venoms was carried out by a number of professional herpetologists, often with little or no reward and in some instances at the ultimate cost of their lives.

Commercial antivenom manufacture began in Australia in 1930 with Tiger Snake antivenom; this was followed later with antivenoms for the other important species:

1930	Tiger Snake
1955	Taipan
1956	Brown snake
1958	Death adder
1959	Black snake
1961	Sea snake
1962	Polyvalent

One of Australia's leading anthropologists and naturalists, Donald Thomson, explored Cape York, Queensland, in 1928, 1929, and 1932–33. During this time, he collected and milked the first specimens of the Coastal Taipan. He assisted Charles Kellaway with research and, in particular, antivenom research. Up until the availability of taipan antivenom in 1955, almost every person bitten by this formidable species died. In 1950 Kevin Clifford Budden, an amateur Australian herpetologist and snake collector, captured a live Coastal Taipan in the Cairns area of northern Queensland for research; in the process, Budden died from a bite from this specimen. His work was instrumental in helping develop a taipan antivenom.

Antivenom is the only specific treatment for snakebite and is produced through first milking snakes of their venom by encouraging them to bite on a rubber diaphragm stretched across a glass beaker. The venom is then dried in a vacuum desiccator and sent to the Commonwealth Serum Laboratories. There it is prepared and sent to a horse farm, where horses are injected with ever-increasing dosages until they have developed sufficient antibodies.

The serum containing the antibodies is then collected and fractioned with immunoglobulins, purified, concentrated, and standardized to contain a minimal number of units. A unit of antivenom will neutralize 0.01 mg of dried venom. The antivenom is then stored away from light and refrigerated at 0–10 °C (it should not be frozen); shelf life is about three years.

Snakebite Treatment: First Aid

Historically it was believed that after a snakebite, venom entered the bloodstream, so the first aid procedure was to wash the bitten area using water, spittle, or urine. The next step was to wash as much venom as possible from the limb with the patient's own blood. This was achieved by taking a razor and making a longitudinal cut along the limb over each puncture wound. If a new razor blade was not available, it was recommended to break a bottle and use a sliver of glass! This archaic method for dealing with snakebite also involved placing a ligature around the bitten limb. The whole procedure was not an effective first aid measure and in fact was quite dangerous for the patient. Venom is not absorbed into the bloodstream from the bite site but moves predominantly through the lymphatic system.

The first aid management of snakebite was revolutionized in 1979 by the work of Dr. Struan Sutherland (1936–2002) and his team in the Department of Immunology Research at the Commonwealth Serum Laboratories in Melbourne. Sutherland's group was responsible for many of the advances in Australian venom research, most notably in the development of the pressure-immobilization technique (described below).

This team of scientists left what is arguably the most important legacy to the field of Australian toxinology. The pressure-immobilization technique was rapidly adopted and has remained the standard first aid care for snakebite victims in Australia. Pressure and immobilization have two functions. Pressure compresses lymphatic drainage to delay the absorption of venom into the microcirculation, and immobilization inhibits gross muscle movement, thereby decreasing intrinsic local pressures that stimulate lymphatic drainage.

After a bite from a dangerous snake, symptoms from envenomation may be experienced. These include general pain, bleeding internally and bruising, muscle paralysis, and difficulty talking, moving, and breathing.

After first aid has been applied, the patient should be hospitalized, and antivenom may need to be administered. A snake venom detection kit (SVDK), developed in 1979, can be

Species	Appropriate antivenom
Death adders	Death adder
Copperheads	Tiger Snake
Tiger Snake	Tiger Snake
Taipans	Taipan
Black snakes	Tiger Snake/Black snake
Brown snakes	Brown snake
Rough-scaled Snake	Tiger Snake
Sea snakes	Sea snake

used as a diagnostic tool to assist in determining the presence of venom and the elapid species responsible for the bite. There have been inconsistencies reported with the use of SVDKs, and while they can aid the choice of antivenom for a patient, a negative result for the presence of venom does not necessarily exclude envenomation.

Australian antivenoms are extremely effective, and specific antivenoms have been produced to neutralize the bites of specific species. An effective polyvalent antivenom is also available for use in cases where the identity of the snake is in question.

Hypersensitivity reactions to antivenom (serum sickness) can occur in patients and may result in fever, a general ill feeling, hives, itching, joint pain, rash, and swollen lymph nodes. Nonsteroidal anti-inflammatory drugs and antihistamines provide symptomatic relief. Severe reactions (multisystem involvement with significant symptoms) may warrant a brief course of corticosteroids.

First Aid for Snakebite: Pressure and Immobilization Method

- Do not wash the bite site.
- Apply a broad pressure bandage from below the bite site, upward on the affected limb. Leave the tips of the fingers or toes unbandaged to allow the patient's circulation to be checked. Do not remove clothing but simply bandage over the clothes.
- Bandage firmly (as for a sprain), tight enough to compress the lymphatic vessels but not enough to prevent circulation.
- If possible, bind a splint to the bitten limb to prevent further movement.
- Apply a pressure pad to bites on the head, neck, or face.
- Keep the patient still, and do not provide food or liquids.
- Seek medical assistance immediately (in Australia call 000).

Snake Conservation

Some people may find the idea of conserving snakes, particularly dangerous snakes, abhorrent, but the role of these reptiles in ecosystems is as important as that of any other form of native wildlife. Certain snake species have become threatened, their population losses attributable to land clearing for agriculture, urban development, and feral animals like foxes and cats impacting the recruitment of juveniles. Maintaining a high level of biodiversity is important, and in Australia, snakes and other reptiles make up a significant proportion of the predators and prey that keep natural ecosystems healthy.

Snake hunt, Murray River, 1906

Since European colonization of Australia, many of the changes made to the environment have led to serious reductions in some snake populations, including venomous snakes.

It can be difficult to accurately assess the processes that threaten snake populations, but certainly land clearing for agriculture and grazing is a serious concern. The alteration of wetlands by removing water from the system and preventing periodic flooding seriously impacts frogs, which are a main food item for many snakes. Clearing, logging, and firewood collection remove vital habitat for snakes. Natural fire regime—the pattern, frequency, and intensity of fires that prevail through an area over a long period of time—is considered an integral part of fire ecology that assists in the renewal of various ecosystems. However, controlled burns, supposedly set to reduce fuel, if too frequent and extensive may have a devastating impact on certain species.

Some species of brown snakes may have benefited from European settlement. They are more common in urban environments and grain-growing areas because of the greater abundance of their primary prey item, the introduced mouse.

Sea snakes are a unique group of marine reptiles found in tropical waters around the world. Global sea snake populations have declined in recent years, including those in the Great Barrier Reef and parts of Western Australia. Declines in native sea snake populations at the remote offshore Ashmore Reef, northwestern Australia, have made these species a focus for conservation and long-term monitoring. While the reasons for decline are not well understood, one key concern in Australian waters is these snakes' frequent capture as bycatch in nets of trawl fisheries. Vulnerability from commercial fishing, pollution, and coastal development are other considerations. Sea snakes are also being studied for responses to increased pressures like rising water temperatures caused by increased global warming.

The conservation status of each species, as determined by the International Union for Conservation of Nature (IUCN) Red List of Threatened Species, is included in the species accounts. Information on recognized threatening processes is also included.

IUCN Conservation Categories

- **Extinct:** no known individuals remaining
- **Critically Endangered:** extremely high risk of extinction in the wild
- **Endangered:** high risk of extinction in the wild
- **Vulnerable:** high risk of endangerment in the wild
- **Near Threatened:** likely to become endangered soon
- **Least Concern:** lowest risk, does not qualify for a more at-risk category; widespread and abundant taxa are included in this category
- **Data Deficient:** not enough data to assess risk of extinction
- **Not Evaluated:** not yet evaluated against the criteria used to assess endangerment

Species Accounts

This guide has been produced to allow readers to identify all the Australian front-fanged venomous snakes (family Elapidae) by using descriptive genus accounts, species accounts, photographs, illustrations, and distribution maps. A glossary provides definitions of unfamiliar terms.

The common and scientific names used in this book follow S. K. Wilson and G. Swan, *A Complete Guide to Reptiles of Australia* (sixth edition, 2021); however, some common names have also been adopted from the Australian Society of Herpetologists' official list of Australian species.

The species accounts are divided into Land Snakes and Marine Snakes and arranged in three categories based on lethality ratings: Dangerously Venomous, Medically Significant, and Potentially Dangerous.

A description of each genus includes the number of Australian species assigned to that genus and a general explanation of morphological features, distribution, and behavior. When known, the type of venom is described and the correct antivenom is listed.

Each individual species is described in greater detail in an account that includes total length (TL, given in meters [m] or millimeters [mm]), morphological features, color and markings, habitat, behavior, reproductive notes, conservation status, and further detailed information to aid in correct identification, including comparisons to other species. Each species account is accompanied by a distribution map and one or more photographs depicting typical coloration and in some cases variation within the species.

Distribution maps are an important tool in helping identify animals, as some species can be quickly ruled out simply based on where they are known to occur. Mapping animals' distribution, or geographical range, is an ongoing challenge, but with the development of the internet and phone cameras, records are being continuously expanded on various databases.

The maps in this guide have been carefully prepared using information from the most current field guides, museum databases, scientific literature, internet databases, and observations from reliable sources. Most species are represented by a map illustrating their general distribution throughout Australia, and some more detailed maps are provided for species with a very limited distribution.

The family Elapidae, distributed throughout the Americas, Africa, Asia, Melanesia, Australia, and the Indian and Pacific Oceans, contains the most venomous species in the world, including the much-feared mambas, cobras, kraits, tiger snakes, and taipans. All of Australia's dangerous snakes are members of this family.

Terrestrial elapid snakes reach their greatest diversity in Australia, with 109 recognized species, three of which are represented by two subspecies each. There are also 32 species (one represented by a subspecies) of marine elapids—sea snakes and sea kraits, which represent two different evolutionary pathways for moving to marine environments from within the family Elapidae.

Elapid snakes have a pair of fixed, immovable, relatively short fangs at the front of the upper jaw. Each is connected at the base to a venom gland via a duct. The venom of Australian elapid snakes is largely neurotoxic, but in some species, it is more hemotoxic, affecting the blood. Features of elapids include enlarged symmetrical scales on top of the head, large ventral scales, and lack of a loreal scale. Some elapid snakes are oviparous (egg-laying), and others viviparous (live-bearing); the latter trait in terrestrial elapids is almost exclusive to Australasian species. Elapids are the largest group of snakes in Australia.

Abbreviations	
ACT	Australian Capital Territory
NSW	New South Wales
NT	Northern Territory
QLD	Queensland
SA	South Australia
TAS	Tasmania
VIC	Victoria
WA	Western Australia

Land Snakes

Dangerously Venomous Land Snakes

This group comprises thirty-three species, one represented by two subspecies, categorized by Mirtschin, Rasmussen, and Weinstein (2017) as capable of delivering a bite to a human that, if untreated, has a high likelihood of a fatal outcome.

Genus *Acanthophis*
Death Adders

A genus of eight broadly accepted species with a very distinctive "viper-like" appearance, *Acanthophis* occurs throughout the Australian continent. Some species also occur in New Guinea and nearby Indonesia. Morphological diagnostics remain problematic, and much further research into identifying defining characteristics is required.

Death adders are cryptic, sedentary snakes that rely on camouflage to ambush prey. They are short and robust, with a broad, triangular head and a thin tail terminating in a soft spine. The eyes are small, with a pale iris and vertically elliptical pupil. Death adders are live-bearing.

Venom is predominantly neurotoxic, weakly hemolytic, cytolytic (attacking the cells), and myotoxic, with weak anticoagulant activity. Death adder or polyvalent antivenom is used to neutralize bites from these species.

Common Death Adder
Acanthophis antarcticus

TL 1 m. **Lethality.** Dangerously venomous. **Description.** Body short and robust with a slender tail terminating in a soft spine. Gray or reddish brown with irregular lighter bands along body. Head triangular, with pointed snout and pale bars on lips. Ventral area creamy gray with darker markings. **Scalation.** Dorsal body scales smooth to weakly keeled and in 21–23 rows at mid-body. Ventrals 110–135. Anal scale single. Subcaudals 35–60, mostly single, some divided posteriorly. **Habitat and range.** Occurs in a variety of habitats from shrublands and heaths to woodlands and rain forests. Distributed in three main areas: through most of e. Australia; from extreme s. SA to s. WA; and in sw. WA. **Behavior.** Diurnal and nocturnal. A well-camouflaged ambush predator, it lies in wait, usually among leaf litter, twitching the tail spine as a lure. Prey items include lizards and small mammals. Litters of up to 33 young recorded. A sedentary snake but capable of delivering swift strikes. **Identification.** *Acanthophis antarcticus* may overlap in range with *A. praelongus* in ne. QLD. The latter species is more slender and has strongly keeled anterior dorsal body scales and supraocular scales forming prominent, raised peaks. **Conservation.** IUCN status: Least Concern. The Common Death Adder has undergone historical declines caused by expanding urbanization and grazing. Inappropriate fire regimes remove ground cover and leaf litter, which this species relies on for refuge and prey items. It is also preyed upon by feral pigs, foxes, and cats.

Common Death Adder (*Acanthophis antarcticus*), red form, Newcastle, NSW

Common Death Adder (*Acanthophis antarcticus*), gray form, Figtree Station, QLD

Common Death Adder (*Acanthophis antarcticus*), Windorah, QLD

Barkly Death Adder
Acanthophis hawkei

TL 1.2 m. **Lethality.** Dangerously venomous. **Description.** Body short and robust with a slender tail terminating in a soft spine. Pale brown to gray with irregular paler, tan to yellowish bands along body. Head triangular, with pointed snout and dark and pale bars on lips. Ventral area plain yellowish. **Scalation.** Dorsal scales smooth to moderately keeled and in 21–23 rows at mid-body. Ventrals 110–155. Anal scale single. Subcaudals 35–60, mostly single, divided posteriorly. **Habitat and range.** Occurs on black-soil plains, grasslands, floodplains, and around swamps, from nw. QLD to south of Darwin in the NT. **Behavior.** Predominantly nocturnal. A well-camouflaged ambush predator, it lies in wait, twitching the tail spine as a lure. Prey items include birds, lizards, frogs, and small mammals. Litters of 8–27 young recorded. A sedentary snake but capable of delivering swift strikes. **Identification.** *Acanthophis hawkei* may overlap in range with *A. rugosus* in nw. QLD. The latter species has prominently keeled dorsal scales anteriorly, supraoculars raised to form prominent peaks, and ventral surface cream with distinct dark blotching. **Conservation.** IUCN status: Vulnerable. Native frogs form much of the diet of the Barkly Death Adder, and it is at risk from eating the toxic introduced Cane Toad (*Rhinella marina*), which has occupied its tropical monsoonal habitat. Also, habitat modification from overgrazing and inappropriate fire regimes are considered potential threats.

Acanthophis hawkei (Fogg Dam, NT)

Smooth-scaled Death Adder
Acanthophis laevis

TL 590 mm. **Lethality.** Dangerously venomous. **Description.** Body short and robust with a slender tail terminating in a soft spine. Coloration highly variable, from gray to green, red, or orange, with irregular darker bands along the body, and tail tip white to yellow. Head triangular, with pointed snout, raised hornlike supraocular scales, and pronounced brow ridge. The lips are often immaculate white, with black blotches on lips and posterior to the eyes and onto the temporals/last supralabials. Ventral area creamy gray. **Scalation.** Dorsal scales smooth and in 21–23 rows at mid-body. Ventrals 110–135. Anal scale single. Subcaudals 35–57, single but divided posteriorly. **Habitat and range.** Occurs in savanna woodlands, kunai (*Imperata cylindrica*) grasslands, rain forests, plantations, and village gardens of mainland New Guinea and the Moluccas in e. Indonesia. In Australia, recorded from Dauan Island, in Torres Strait. **Behavior.** Typically nocturnal. A well-camouflaged ambush predator, it lies in wait, usually among ground debris, twitching the tail spine as a lure. Prey items include lizards, frogs, birds, and small mammals. Litters of up to 20 young recorded. A sedentary snake but capable of delivering swift strikes. **Identification.** *Acanthophis laevis* may overlap in distribution with *A. praelongus* on Dauan Island. It is distinguished from the latter species by its distinctive white lips with notable black blotches rather than darker bars; *A. laevis* also has the most highly raised supraoculars of all the death adder species. **Conservation.** IUCN status: Least Concern throughout its Papuan New Guinea and Indonesian range; not listed for Australia.

Smooth-scaled Death Adder (*Acanthophis laevis*), Martin River, Central Province, Papua New Guinea

Kimberley Death Adder
Acanthophis lancasteri

TL 645 mm. **Lethality.** Dangerously venomous. **Description.** Body short and robust with a slender tail terminating in a soft spine. Dull orange, tan, or gray with darker irregular bands along the body and a black tail tip. Head triangular, with pointed snout, raised supraoculars forming peaks, and lips pale with dark mottling. Ventral area cream to yellow. **Scalation.** Dorsal scales prominently keeled and in 22–23 rows at mid-body. Ventrals 125–139. Anal scale divided. Subcaudals 46–56, mostly single but divided posteriorly. **Habitat and range.** Restricted to the n. Kimberley region of n. WA, where it has been recorded from savanna woodlands with grass tussocks, among basalt boulders, along creek banks, and beneath rocks in vine thickets. **Behavior.** Nocturnal. A well-camouflaged ambush predator, it lies in wait, twitching the tail spine as a lure. Prey items are presumably lizards, frogs, and small mammals. Litters of up to 27 young recorded. A sedentary snake but capable of delivering swift strikes. **Identification.** *Acanthophis lancasteri* is the only death adder species in the n. Kimberley region. Its range abuts with that of *A. rugosus* around Keep River, NT. **Conservation.** IUCN status: Vulnerable. With the recent arrival of the introduced Cane Toad in the habitat of this species, it is probably at risk of a significant decline.

Kimberley Death Adder (*Acanthophis lancasteri*), Kimberley, WA

Northern Death Adder
Acanthophis praelongus

TL 600 mm. **Lethality.** Dangerously venomous. **Description.** Body relatively slender with a slender tail terminating in a soft spine. Gray to reddish brown or yellow, with irregular lighter bands along the body. Head triangular, with pointed snout and white lips with darker bars. Raised supraoculars form very prominent peaks. Ventral area creamy-colored. **Scalation.** Dorsal scales strongly keeled anteriorly and in 21–23 rows at mid-body. Ventrals 110–135. Anal scale single. Subcaudals 35–60, mostly single, some divided posteriorly. **Habitat and range.** Occurs in heaths, eucalypt woodlands, and vine thickets, and among leaf litter in rock outcrops, from about Townsville north through Cape York Peninsula, QLD, and Torres Strait Islands. **Behavior.** Nocturnal and cryptic. A well-camouflaged ambush predator, it lies in wait, usually in leaf litter, twitching the tail spine as a lure. Prey items include lizards and small mammals. Litters of 6–17 young recorded. A sedentary snake but capable of delivering swift strikes. **Identification.** *Acanthophis praelongus* may overlap with *A. antarcticus* in ne. QLD. The latter species is more robust, has weakly keeled dorsal body scales, and the supraocular scales do not form prominent raised peaks. On Dauan Island in Torres Strait, *A. praelongus* also may occur with *A. laevis*, which has distinctive white lips with notable black blotches rather than bars. **Conservation.** IUCN status: Least Concern. The main threat to this species is inappropriate fire regimes that destroy leaf-litter substrate. Grazing and expanding pastoralism are also concerns. The consumption of Cane Toads may represent some threat, but this species has coexisted with toads for decades.

Northern Death Adder (*Acanthophis praelongus*), Iron Range, QLD

Desert Death Adder
Acanthophis pyrrhus

TL 710 mm. **Lethality.** Dangerously venomous. **Description.** Body relatively slender with a slender tail terminating in a soft spine. Pale reddish brown to darker red with narrow, irregular cream to yellow bands along the body. Head triangular, with pointed snout and mottled lips. Ventral area cream to white. **Scalation.** Dorsal and lateral scales strongly keeled and usually in 21 rows at mid-body. Ventrals 120–162. Anal scale single. Subcaudals 45–67, mostly single, a few divided posteriorly. **Habitat and range.** Occurs in arid sand plains and stony deserts in association with spinifex and acacia scrublands, from sw. QLD across central Australia to north and south of the Pilbara region of WA. **Behavior.** Mostly nocturnal and crepuscular. A well-camouflaged ambush predator, it half buries itself in sand or soil, twitching the tail spine as a lure. Prey items include lizards and small mammals. Litters of 9–14 young recorded. A sedentary snake but capable of delivering swift strikes. **Identification.** *Acanthophis pyrrhus* may overlap with *A. wellsi* on the edge of the Pilbara region. The latter species has body scales moderately keeled dorsally and smooth laterally. **Conservation.** IUCN status: Least Concern. Fire represents a local threat, but foxes and cats are a bigger concern as they prey on this species and its prey items.

Desert Death Adder (*Acanthophis pyrrhus*), Port Hedland, WA.
Insert: Desert Death Adder (*Acanthophis pyrrhus*), head, Port Hedland, WA.

Top End Death Adder
Acanthophis rugosus

TL 700 mm. **Lethality.** Dangerously venomous. **Description.** Body short and robust with a slender tail terminating in a soft spine. Gray to reddish brown with irregular darker bands along the body. Head triangular, with pointed snout, distinctly mottled or barred upper lip, and raised supraoculars forming prominent peaks. Ventral area creamy-colored with distinct dark blotching. **Scalation.** Dorsal scales prominently keeled anteriorly and in 21–23 rows at mid-body. Ventrals 115–165. Anal scale single. Subcaudals 53, mostly single but divided posteriorly. **Habitat and range.** Associated with monsoon forest habitats, savanna woodlands, grasslands, stony hills, and outcrops, from Mt. Isa area of nw. QLD and through the Top End of the NT. Also occurs in s. New Guinea and nearby Indonesia. **Behavior.** Diurnal and nocturnal. A well-camouflaged ambush predator, it lies in wait, usually twitching the tail spine as a lure. Prey items include lizards and small mammals. Litters of 6–24 young recorded. A sedentary snake but capable of delivering swift strikes. **Identification.** *Acanthophis rugosus* may overlap in range with *A. hawkei* in nw. QLD. The latter species has smooth to moderately keeled dorsal scales and ventral surface pale yellow with some lighter flecking; it also lacks raised supraoculars forming prominent peaks. **Conservation.** IUCN status: Least Concern. In some areas of Australia, Cane Toads have caused serious declines of this species.

Top End Death Adder (*Acanthophis rugosus*), John Hills, Winton, QLD

Pilbara Death Adder
Acanthophis wellsi

TL 520 mm. **Lethality.** Dangerously venomous. **Description.** Body relatively slender with a slender tail terminating in a soft spine. Three distinct color forms: red with black head and black bands; pale brown to orange with irregular lighter bands along body; pale overall. Head triangular, with pointed snout, and lips generally not barred. Ventral area light with darker flecking. **Scalation.** Moderately keeled dorsally, smooth laterally, and in 21 rows at mid-body. Ventrals 119–143. Anal scale single. Subcaudals 41–64, mostly single but divided posteriorly. **Habitat and range.** Occurs in association with stony ranges and sandy areas with spinifex and acacia scrublands in WA; widespread in the Pilbara region, and an isolated population at North West Cape. **Behavior.** Diurnal and nocturnal. A well-camouflaged ambush predator, it half buries itself, twitching the tail spine as a lure. Prey items are mostly skinks and small mammals. Litters of 9–20 young recorded. A sedentary snake but capable of delivering swift strikes. **Identification.** *Acanthophis wellsi* is largely restricted to the Pilbara region and has body scales moderately keeled dorsally and smooth laterally. *Acanthophis pyrrhus* occurs in close proximity and has strongly keeled dorsal and lateral scales. The two species may hybridize close to the North West Cape. **Conservation.** IUCN status: Least Concern. No threats listed.

Pilbara Death Adder (*Acanthophis wellsi*), pale overall, Pilbara, WA

Pilbara Death Adder (*Acanthophis wellsi*), pale brown to orange with irregular lighter bands along body, Mt. Sheila, WA

Pilbara Death Adder (*Acanthophis wellsi*), red with black head and black bands, Mt. Sheila, WA

Genus *Austrelaps*
Copperheads

The three species in this genus have a preference for cooler environments and occur in different areas of se. Australia, including TAS and some islands in Bass Strait. Medium-size to large snakes, copperheads have barred lips, moderately large eyes, a pale iris, and a round pupil. Prey items include frogs, lizards, and small mammals. These snakes are live-bearing.

Venom is neurotoxic, hemolytic, and cytotoxic, with weak anticoagulant activity. Tiger Snake or polyvalent antivenom is used to neutralize bites from these species.

Pygmy Copperhead
Austrelaps labialis

TL 870 mm. **Lethality.** Dangerously venomous. **Description.** Body medium-size and robust with a short tail. Pale brown to dark gray, sometimes with a darker vertebral stripe. Head with moderately pointed snout; often has dark bar across back of head; lips prominently barred. Ventral area cream to yellow, sometimes with orange flecks. **Scalation.** Dorsal scales smooth and weakly glossed in 15 (rarely 17) rows at mid-body. Ventrals 133–155. Anal scale single. Subcaudals 35–58, all single. **Habitat and range.** Restricted to cooler moist habitats in grasslands, swamps, and wetlands within open forests and woodlands in the Mt. Lofty Ranges, base of the Fleurieu Peninsula, and on Kangaroo Island, SA. **Behavior.** Mostly diurnal but crepuscular in hot weather. May be observed foraging throughout the day or sheltering beneath rocks, logs, and other ground debris. Active at low temperatures. Feeds primarily on frogs and lizards. Varying litters of 3–32 young recorded. An inoffensive species not inclined to bite. **Identification.** *Austrelaps labialis* is the only copperhead species within its range. **Conservation.** IUCN status: Least Concern. No threats listed.

Pygmy Copperhead (*Austrelaps labialis*), Kangaroo Island, SA

Highlands Copperhead
Austrelaps ramsayi

TL 1.1 m. **Lethality.** Dangerously venomous. **Description.** Body moderately large and robust with a short tail. Pale gray to reddish brown to black, with enlarged lighter cream, yellow, or red lateral scales. Head with moderately pointed snout and prominently white-edged, barred lips. Ventral area cream to gray. **Scalation.**Dorsal scales smooth, weakly glossed, and in 15 (rarely 17) rows at mid-body. Ventrals 150–170. Anal scale single. Subcaudals 35–55, all single. **Habitat and range.** Occurs in dry and wet sclerophyll forests, woodlands, and heathlands in cooler upland areas of se. Australia. **Behavior.** Mostly diurnal. Observed foraging throughout the day or sheltering beneath rocks, logs, and other ground debris. Active at low temperatures. Feeds primarily on frogs and lizards. Litters of up to 20 young recorded. An inoffensive species not inclined to bite. **Identification.** *Austrelaps ramsayi* may overlap in distribution with *A. superbus*, but the former occurs at higher elevations and is distinguished by its smaller adult size and usually more prominently barred lips. **Conservation.** IUCN status: Least Concern. Existing threats are localized and include the draining of swamps and other wetlands.

Highlands Copperhead (*Austrelaps ramsayi*), Taralga, NSW

Lowlands Copperhead
Austrelaps superbus

TL 1.8 m. **Lethality.** Dangerously venomous. **Description.** Body very large and robust with a short tail. Light gray to reddish brown to black, with enlarged lighter cream, yellow, pink, or orange lateral scales. Head with moderately pointed snout and white-edged, weakly barred lips; sometimes with a dark or light band across nape. Ventral area cream to gray. **Scalation.** Dorsal scales smooth, weakly glossed, and in 15 (rarely 17) rows at mid-body. Ventrals 140–165. Anal scale single. Subcaudals 35–55, all single. **Habitat and range.** The favored habitat is tussock-edged swamps in dry sclerophyll forests and woodlands in low-lying areas of se. SA, VIC, TAS, and some Bass Strait islands. **Behavior.** Mostly diurnal. Usually observed basking during the day or sheltering beneath rocks, logs, and other ground debris. Active at low temperatures. Feeds primarily on frogs and lizards. Litters of 2–32 young recorded. A relatively shy snake, it quickly seeks shelter if disturbed. **Identification.** *Austrelaps superbus* may overlap in distribution with *A. ramsayi* but is distinguished from the latter species by its more robust, larger adult size and its weakly barred lips. **Conservation.** IUCN status: Least Concern. No threats listed.

Lowlands Copperhead (*Austrelaps superbus*), Campbell Town, TAS

Lowlands Copperhead (*Austrelaps superbus*), males combatting, Seaford, VIC

Genus *Notechis*
Tiger Snake

Until very recently, this genus was considered to contain a single species with six subspecies. A detailed molecular study demonstrated that, across the genus's range, all populations of *Notechis* are genetically extremely closely related, and their morphological differences evolved very recently. *Notechis scutatus* is therefore considered a single, morphologically variable species. It occurs in separate populations in sw. Australia, SA, and e. Australia as far north as se. QLD, as well as TAS and various islands. Often found in association with water, it is a large snake with a broad, relatively flat head and moderately large eyes. Prey items include frogs, lizards, small mammals, and birds. The Tiger Snake is live-bearing.

Venom is geographically variable and is strongly neurotoxic, procoagulant, weakly hemolytic, and cytotoxic. Tiger Snake or polyvalent antivenom is used to neutralize bites from this species.

Tiger Snake
Notechis scutatus

TL 1.7 m (Eastern mainland population); TL 900 mm (Flinders Ranges population); TL 2 m (Tasmanian population); TL 1.2m (SA peninsula population); TL 1.6 m (WA population); TL 2.1 m (Mt. Chappell Island population). **Lethality.** Dangerously venomous. **Description.** Body medium-size to large and robust with a short, slender tail. Head broad, slightly distinct from neck, with blunt snout. Ventral area cream, yellow, gray, or olive green. **Scalation.** Dorsal scales smooth and in 17 rows at mid-body. Ventrals 140–190. Anal scale single. Subcaudals 35–65, all single. EASTERN MAINLAND POPULATION: Coloration highly variable but usually pale gray or brown to olive green with a series of narrow crossbands formed by pale-edged scales. Some individuals lack banding entirely and are a uniform color. Black individuals are recorded from various areas in se. Australia. FLINDERS RANGES POPULATION: Generally uniform black but sometimes with a series of pale, narrow crossbands. TASMANIAN POPULATION: Coloration highly variable, from jet black to pale yellow with gray bands, yellow, or various shades of brown with or without some lighter banding. SA PENINSULA POPULATION: Usually plain black or coppery brown, sometimes with a series of narrow, pale crossbands. WA POPULATION: Generally black or brown with a series of narrow, bright yellow crossbands and lower flanks yellow, orange, or whitish. MT. CHAPPELL ISLAND POPULATION: Usually black, dark brown, or yellowish, sometimes with narrow, paler crossbands. **Habitat and range.** EASTERN MAINLAND POPULATION: Variety of habitats from rain forests in the northeast to dry sclerophyll forests, marshlands, river floodplains, and developed areas of se. Australia. FLINDERS RANGES POPULATION: Along small creeks in dry open forests and shrublands from the s. Flinders Ranges to Mt. Remarkable National Park in SA. TASMANIAN POPULATION: Variety of habitats including rain forests, heaths, wetlands, and developed areas throughout TAS; also occurs on King Island and other smaller islands of w. Bass Strait, where it occupies shearwater burrows. SA PENINSULA POPULATION: Forests, grasslands, heaths, and shrublands in s. SA on Kangaroo Island, w. Eyre Peninsula, s. Yorke Peninsula, and some associated offshore islands. WA POPULATION: Wetlands, scrublands, and developed areas in sw. WA; also found on Garden and Carnac Islands. MT. CHAPPELL ISLAND POPULATION: Grasslands and heaths of the Furneaux Group of islands in e. Bass Strait, including Flinders and Mt. Chappell Islands. **Behavior.** Diurnal and crepuscular but also nocturnal in warm weather. A large, active snake generally (apart from island populations) encountered around waterways. It shelters beneath ground debris and is relatively inoffensive. Diet consists of frogs, lizards, birds, and small mammals. Average litters of 23 young recorded. **Identification.** Notechis scutatus is a moderate-size to large, distinctive species with variable patterns throughout its range. In the northeastern part of its range (n. NSW and se. QLD), the eastern mainland population could be confused with the Rough-scaled Snake (*Tropidechis carinatus*), but the latter species has keeled scales. **Conservation.** IUCN status: Least Concern. EASTERN MAINLAND POPULATION: Considered to have declined through loss of habitat; potential predators include cats, foxes, and dogs; in northeastern part of range, a threat exists from eating Cane Toads. FLINDERS RANGES POPULATION: Threats include overgrazing, land clearing, soil erosion, water pollution, and inappropriate fire regimes; introduced trout compete for prey items. WA POPULATION: Large-scale development on the Swan Coastal Plain wetlands, WA, has reduced the size of local populations. TASMANIAN, SA PENINSULA, AND MT. CHAPPELL ISLAND POPULATIONS: No threats listed.

Eastern mainland population

Tiger Snake (*Notechis scutatus*), Olney State Forest, NSW

Tiger Snake (*Notechis scutatus*), Lake Alexandrina, SA

Tiger Snake (*Notechis scutatus*), unbanded form, Stony Rises, VIC

Flinders Ranges population

Tiger Snake (*Notechis scutatus*), Flinders Ranges form, Southern Flinders Ranges, SA

Tasmanian population

Tiger Snake (*Notechis scutatus*), Campbell Town, TAS

Tiger Snake (*Notechis scutatus*), Tasmanian form, close-up, TAS

SA peninsula population

Tiger Snake (*Notechis scutatus*), peninsular form, Kangaroo Island, SA

WA population

Tiger Snake (*Notechis scutatus*), western form, Perth, WA

Tiger Snake (*Notechis scutatus*), western form, head, Mandurah, WA

Mt. Chappell Island
population

Tiger Snake (*Notechis scutatus*), Chappell Island form, Chappell Island, TAS

Genus *Oxyuranus*
Taipans

Australia's largest venomous snakes, taipans comprise three distinct species, one represented by a mainland subspecies and a poorly defined island subspecies. They occur in tropical woodlands in separate regions of n. and ne. Australia, including the Torres Strait Islands, and New Guinea; in arid plains of the e. interior; and in the remote w. deserts of WA and the NT. Taipans are large and dangerous terrestrial snakes with long narrow heads, large eyes, and round pupils with a dark to pale iris. Prey items are exclusively mammals, particularly rats. Taipans are oviparous. Similar species include the brown snakes (genus *Pseudonaja*).

Taipans have long fangs, and the venom is strongly neurotoxic with procoagulant properties and weakly hemolytic with cytotoxic and myotoxic activities. Taipan or polyvalent antivenom is used to neutralize bites from these species.

Inland Taipan
Oxyuranus microlepidotus

TL 2 m. **Lethality.** Dangerously venomous. **Description.** Body large and robust with a slender tail. Light brown to rich brown with dark-edged scales that form irregular crossbands. Undergoes a seasonal color change from pale in the summer to dark in the winter. Head long and narrow, slightly distinct from neck, with narrowly rounded snout. Ventral area cream to bright yellow. **Scalation.** Dorsal scales smooth and in 23 rows at mid-body. Ventrals 220–250. Anal scale single. Subcaudals 55–70, all divided. **Habitat and range.** Associated with black-soil plains and floodplains of the Lake Eyre basin. A few very old records, possibly erroneous, exist from adjacent areas. **Behavior.** Diurnal. An active species that shelters in deep cracks in the soil and in burrow systems. Prey items are mostly rats; the main food source is the Long-haired Rat (*Rattus villosissimus*). Average clutches of 16 eggs recorded. Raises its forebody in an S-shaped position if disturbed. Considered to be the most venomous land snake in the world. **Identification.** *Oxyuranus microlepidotus* is a large, distinctive species. Most likely to be confused with the Strap-snouted Brown Snake (*Pseudonaja aspidorhyncha*) or the Western Brown Snake (*P. mengdeni*), but those species have 17 mid-body scale rows and a divided anal scale. **Conservation.** IUCN status: Least Concern. Threats may include competition for its main food source from foxes and cats and also, possibly, the development of large-scale irrigation projects.

Inland Taipan (*Oxyuranus microlepidotus*), central SA

Inland Taipan (*Oxyuranus microlepidotus*), summer coloration, Coober Pedy, SA

Inland Taipan (*Oxyuranus microlepidotus*), winter coloration, Durham Downs Station, QLD

Coastal Taipan
Oxyuranus scutellatus scutellatus

TL 3 m. **Lethality.** Dangerously venomous. **Description.** Body large and robust with a slender tail. Coloration variable, ranging from dark brown to yellow, gray, or black. Head pale, long and narrow, slightly distinct from neck, with narrowly rounded snout. Ventral area yellow to orange, sometimes with red flecking. **Scalation.** Dorsal scales smooth and in 21–23 rows at mid-body. Ventrals 220–250. Anal scale single. Subcaudals 45–80, all divided. **Habitat and range.** Tropical woodlands, savanna grasslands, rain-forest verges, cane fields, and dry forests from coastal regions of n. NSW to ne. QLD and across the Top End of the NT to the Kimberley region of WA. **Behavior.** Diurnal. An active snake often observed on well-timbered grassy slopes. Shelters in burrows. Diet consists of mammals, particularly rats. Clutches of 5–17 eggs recorded. A formidable species with a nervous disposition, capable of delivering rapid bites in quick succession. **Identification.** *Oxyuranus scutellatus scutellatus* is a large, distinctive subspecies. May possibly be confused with the Eastern Brown Snake (*Pseudonaja textilis*) or the Northern Brown Snake (*P. nuchalis*), but those species have 17 mid-body scale rows and a divided anal scale. Also similar in appearance to the Greater Black Whipsnake (*Demansia papuensis*), which has only 15 mid-body scale rows. **Conservation.** IUCN status: Least Concern. No threats listed.

Coastal Taipan (*Oxyuranus scutellatus scutellatus*), Cooktown, QLD. Insert: Coastal Taipan (*Oxyuranus scutellatus scutellatus*), close-up, Cooktown, QLD.

Coastal Taipan (*Oxyuranus scutellatus scutellatus*), Cooktown, QLD

Coastal Taipan (*Oxyuranus scutellatus scutellatus*), hatching

Papuan Taipan
Oxyuranus scutellatus canni

TL 3 m. **Lethality.** Dangerously venomous. **Description.** Body large and robust with a slender tail. Coloration variable but usually dark brown or black with a distinct reddish-or-ange vertebral stripe, widening posteriorly. Head pale, long and narrow, slightly distinct from neck, with narrowly rounded snout. Ventral area yellow to orange, sometimes with red flecking. **Scalation.** Dorsal scales smooth and in 21–23 rows at mid-body. Ventrals 220– 250. Anal scale single. Subcaudals 45–80, all divided. **Habitat and range.** Tropical woodlands, savanna grass-lands, rain-forest verges, cane fields, and dry forests. Widespread in New Guinea but in Australia known only from some Torres Strait islands. **Behavior.** Diurnal. An active snake usually observed foraging or basking. Shelters in burrows. Diet consists of mammals, particularly rats. Clutches of 5–17 eggs recorded. A formida-ble species with a nervous disposition, capable of delivering rapid bites in quick succession. **Identification.** *Oxyuranus scutellatus canni* was originally described, with weak characteristics, in 1956 from New Guinea and offshore islands. Several studies with molecular genetic analysis suggest it should be synonymized with *O. scutellatus scutellatus*. May possibly be confused with the Eastern Brown Snake (*Pseudonaja textilis*), but that species has 17 mid-body scale rows and a divided anal scale. On some Torres Strait islands, *O. s. canni* could be confused with the Papuan Black Snake (*Pseudechis papuanus*), but that species has a heavier build, broader depressed head, and a divided anal scale. **Conservation.** IUCN status: Least Concern. No threats listed.

Papuan Taipan (*Oxyuranus scutellatus canni*), Sabai Island, QLD

Western Desert Taipan
Oxyuranus temporalis

TL 1.7 m. **Lethality.** Dangerously venomous. **Description.** Body large and robust with a slender tail. Yellow to light or dark brown. Undergoes a seasonal color change from pale in the summer to dark in the winter. Head pale, long and narrow, slightly distinct from neck, with narrowly rounded snout. Ventral area cream to yellow with small orange spots anteriorly. **Scalation.** Dorsal scales smooth and in 21 rows at mid-body. Ventrals 240–252. Anal scale single. Subcaudals 56–61, all divided. **Habitat and range.** Associated with red sand and dune-field habitats with scattered vegetation. There are records from the eastern edge of the Walter James Range and the Great Victoria Desert of WA and from the George Gill Range (an older record) and other areas (more recently) of the NT. This species undoubtedly has a larger distribution than is currently recognized. **Behavior.** Diurnal. Probably shelters in burrow systems. Diet consists of small mammals. Clutch size not known. The species is recently described, and little is known of its habits and behavior. **Identification.** *Oxyuranus temporalis* is a large, distinctive species, most likely to be confused with the Western Brown Snake (*Pseudonaja mengdeni*), but that species has 17 mid-body scale rows and a divided anal scale. **Conservation.** IUCN status: Least Concern. No threats listed.

Western Desert Taipan (*Oxyuranus temporalis*), Ilkurlka Roadhouse, WA

Genus *Pseudechis*
"Black" Snakes

This genus of nine medium-size to large, robust snakes has eight species distributed throughout Australia; New Guinea shares one of these species and is home to another. The genus contains a dwarf lineage, and genetic evidence suggests there may be more species than are currently recognized. Generally referred to as "black" snakes, though most are not black, these snakes have a relatively broad, depressed head and moderately small eyes with round pupils. Prey items include fishes, frogs, reptiles, and mammals. Most *Pseudechis* species are oviparous, and one is live-bearing.

Venom is strongly myotoxic and hemolytic with strong anticoagulant actions. It is also cytotoxic and mildly neurotoxic. Tiger Snake, black snake, or polyvalent antivenom is used to neutralize bites from these species.

Mulga Snake or King Brown Snake
Pseudechis australis

TL 2.5 m. **Lethality.** Dangerously venomous. **Description.** Body large and robust with a slender tail. Coloration variable, ranging from black to pale yellow, rich copper, or reddish brown dorsally, with lighter ventral coloration extending up onto the lower flanks. Head relatively broad and depressed, slightly distinct from neck, with moderately rounded snout. Ventral area cream to white with scattered orange blotches and some salmon markings. **Scalation.** Dorsal scales smooth, glossy, and in 17 rows at mid-body. Ventrals 185–225. Anal scale divided. Subcaudals 50–80, usually single with last few divided. **Habitat and range.** Widespread through tropical forests, floodplains, semiarid scrublands, grasslands, and deserts, excluding parts of s. and se. Australia. **Behavior.** Diurnal and nocturnal. One of Australia's largest venomous snakes. An active species usually observed foraging. Shelters in burrows and beneath ground debris. Prey items include small mammals, lizards, frogs, and other snakes. Average clutches of 9 eggs recorded. Capable of expressing large quantities of venom. **Identification.** *Pseudechis australis* is a large, distinctive species but may be confused with the pygmy mulga snakes (*P. pailsei* and *P. weigeli*), which attain a smaller adult size and have a single anal scale. Brown snakes (genus *Pseudonaja*) may be similar but have all subcaudal scales divided. **Conservation.** IUCN status: Least Concern. This species has suffered serious declines in areas that Cane Toads have invaded.

Mulga Snake/King Brown Snake (*Pseudechis australis*), Southern Brigalow, QLD

Mulga Snake/King Brown Snake (*Pseudechis australis*), Almaden, QLD

Spotted Mulga Snake
Pseudechis butleri

TL 1.6 m. **Lethality.** Dangerously venomous. **Description.** Body large and robust with a slender tail. Black to dark brown with cream to yellow spots. Head relatively broad and depressed, slightly distinct from neck, with moderately rounded snout. Ventral area yellow, with dark edges on scales and sometimes black flecking. **Scalation.** Dorsal scales smooth, glossy, and in 17 rows at mid-body. Ventrals 200–225. Anal scale divided. Subcaudals 50–70, usually single with last few divided. **Habitat and range.** Occurs in arid environments with mulga and acacia woodlands, generally on stony soils of the Goldfields region of s. inland WA. **Behavior.** Diurnal and nocturnal. An active species usually observed foraging. Shelters in burrows. Diet consists of lizards, other snakes, and small mammals. Clutches of 7–17 eggs recorded. **Identification.** The distinctive coloring separates *Pseudechis butleri* from *P. australis*, with which it is sympatric. **Conservation.** IUCN status: Least Concern. No threats listed.

Spotted Mulga Snake (*Pseudechis butleri*), Yalgoo, WA

Collett's Snake
Pseudechis colletti

TL 1.5 m. **Lethality.** Dangerously venomous. **Description.** Body large and robust with a slender tail. Black or gray to dark reddish brown with multiple irregular cream, pink, or red bands and spots, and a pink to orange flush on the lower flanks. Head relatively broad and depressed, slightly distinct from neck, with moderately rounded snout. Ventral area orange, usually with darker flecking. **Scalation.** Dorsal scales smooth, glossy, and in 19 rows at mid-body. Ventrals 215–235. Anal scale divided. Subcaudals 45–65, usually single with last few divided. **Habitat and range.** Restricted to deeply cracking clay-soil plains with Mitchell grass (*Astrebla*) in the arid interior of QLD. **Behavior.** Diurnal and nocturnal. An active species that shelters in burrows and crevices in deeply cracking soil. Prey items include frogs, lizards, snakes, and small mammals. Clutches of 7–18 eggs recorded. **Identification.** *Pseudechis colletti* is a distinctively colored snake with a limited distribution and unlikely to be confused with other species. **Conservation.** IUCN status: Least Concern. Possible threats include ingestion of Cane Toads and habitat degradation through overgrazing.

Collett's Snake (*Pseudechis colletti*), Winton, QLD

Collett's Snake (*Pseudechis colletti*), juvenile, Winton, QLD

Spotted Black Snake
Pseudechis guttatus

TL 1.5 m. **Lethality.** Dangerously venomous. **Description.** Body large and robust with slender tail. Glossy black to gray, often with scattered pale gray or cream spots. Head relatively broad and depressed, slightly distinct from neck, with moderately rounded snout. Ventral area gray or blue-gray. **Scalation.** Dorsal scales smooth, glossy, and in 19 rows at mid-body. Ventrals 175–205. Anal scale divided. Subcaudals 45–65, usually single with last few divided. **Habitat and range.** Associated mostly with river floodplains, dry sclerophyll forests, grasslands, and woodlands in ne. NSW and se. QLD. **Behavior.** Diurnal and nocturnal. An active species usually observed basking or foraging. Shelters in burrows and cracking soil. Prey items include frogs, lizards, snakes, and small mammals. Clutches of 5–17 eggs recorded. **Identification.** *Pseudechis guttatus* is a large, distinctively colored species recognizable by the glossy black to gray dorsum and gray or blue-gray ventral surface. **Conservation.** IUCN status: Least Concern. Some populations of this species are possibly under threat from habitat degradation through plowing for heavy agricultural use.

Spotted Black Snake (Pseudechis guttatus), Inglewood, QLD

Spotted Black Snake (Pseudechis guttatus), Oakey, QLD

Spotted Black Snake (Pseudechis guttatus), Sundown National Park, QLD

Eastern Pygmy Mulga Snake
Pseudechis pailsei

TL 1.2 m. **Lethality.** Dangerously venomous. **Description.** Body moderately large and slender with a slender tail. Pale yellow, with darker flecks on neck. Head relatively narrow, slightly distinct from neck, with moderately rounded snout. Ventral area yellow to cream. **Scalation.** Dorsal scales smooth, glossy, and in 17 rows at mid-body. Ventrals 210–235. Anal scale single. Subcaudals 50–80, all single. **Habitat and range.** Habitat and range. Occurs in arid scrublands with rock and spinifex, from near Winton in w. QLD almost to the ne. NT. It appears that an undescribed species may occur throughout most of the Top End. **Behavior.** Diurnal and nocturnal. Usually observed foraging. Shelters in burrows and beneath ground debris. Prey items include frogs, lizards, and small mammals. Clutches of 5–11 eggs recorded. **Identification.** *Pseudechis pailsei* is distinguished from the larger *P. australis* by its smaller adult size, narrow head, single subcaudal scales, and single anal scale. May be confused with brown snakes (genus *Pseudonaja*), but those species have all subcaudal scales divided. **Conservation.** IUCN status: Least Concern. No threats listed.

Eastern Pygmy Mulga Snake (*Pseudechis pailsei*), John Hills, Winton, QLD

Papuan Black Snake
Pseudechis papuanus

TL 2.1 m. **Lethality.** Dangerously venomous. **Description.** Body large and robust with a slender tail. Black to dark gray or brown, sometimes with a reddish ventral flush. Head relatively broad and depressed, slightly distinct from neck, with moderately rounded snout. Ventral area dark gray. **Scalation.** Dorsal scales smooth, glossy or matte, and in 19–21 rows at midbody. Ventrals 205–230. Anal scale divided. Subcaudals 49–63, usually single with last few divided. **Habitat and range.** Occurs in swamps, woodlands, and savanna grasslands in low-lying areas of s. New Guinea. In Australia recorded only from Boigu and Saibai Islands in Torres Strait. **Behavior.** Diurnal and nocturnal. An active species usually observed foraging. Shelters in burrows and beneath ground debris. Prey items include frogs, lizards, snakes, and small mammals. Clutches of 7–18 eggs recorded. Considered to be the most toxic of the snakes in this genus. **Identification.** *Pseudechis papuanus* is a large, usually black snake that has been confused with the Papuan Taipan (*Oxyuranus scutellatus canni*). It is distinguished from that species by its heavier build, broader depressed head, and divided anal scale. **Conservation.** IUCN status: Data Deficient. More research is required to determine distribution, population status, and threats.

Papuan Black Snake (*Pseudechis papuanus*), Sabai Island, QLD

Red-bellied Black Snake
Pseudechis porphyriacus

TL 2 m. **Lethality.** Dangerously venomous. **Description.** Body large and robust with a slender tail. Glossy black with usually red (sometimes cream) ventral coloration extending up on the lower flanks. Head relatively broad and depressed, slightly distinct from neck, with a moderately rounded snout. Ventral area red to cream. **Scalation.** Dorsal scales smooth, glossy, and in 17 rows at mid-body. Ventrals 170–215. Anal scale divided. Subcaudals 40–65, usually single with last few divided. **Habitat and range.** Found mostly in association with wetlands in e. Australia, from SA, VIC, and NSW to QLD. **Behavior.** Diurnal. A familiar species generally observed basking or foraging. Shelters beneath logs and other ground debris. Prey items include frogs, lizards, snakes, and small mammals. Litters of 15–18 young recorded, born in clear membranous sacs from which they quickly emerge. **Identification.** *Pseudechis porphyriacus* is a large, distinctively colored species recognizable by its glossy black dorsum and red to cream ventral surface. Smaller specimens could be confused with the Eastern Small-eyed Snake (*Cryptophis nigrescens*), which has only 15 mid-body scale rows and lacks the continuation of red ventral pigment up to the lower lateral zone. **Conservation.** IUCN status: Least Concern. Consumption of Cane Toads by this species is considered to have caused local population declines.

Red-bellied Black Snake (*Pseudechis porphyriacus*), Sydney, NSW

Red-bellied Black Snake (*Pseudechis porphyriacus*), males combating, Tuross Head, NSW

Western Pygmy Mulga Snake
Pseudechis weigeli

TL 1.2 m. **Lethality.** Dangerously venomous. **Description.** Body moderately large and slender with a slender tail. Pale yellow to brown or gray, with darker flecks on the neck. Western populations have a more reticulated pattern formed by paler scale bases. Head relatively narrow, slightly distinct from neck, with moderately rounded snout. Ventral area yellow to cream. **Scalation.** Dorsal scales smooth, glossy, and in 17 rows at mid-body. Ventrals 210–230. Anal scale single. Subcaudals 50–80, all single. **Habitat and range.** Occurs in dry woodlands, grasslands, and rocky outcrops from the Top End of the NT to the Kimberley region of WA. **Behavior.** Diurnal and nocturnal. Usually observed foraging. Shelters in burrows and beneath ground debris. Prey items include frogs, lizards, and small mammals. Clutches of 7–12 eggs recorded. **Identification.** *Pseudechis weigeli* is distinguished from the larger *P. australis* by its smaller adult size, narrow head, single subcaudal scales, and single anal scale. May be confused with brown snakes (genus *Pseudonaja*), but those species have all subcaudal scales divided. **Conservation.** IUCN status: Least Concern. Possibly under threat through the ingestion of Cane Toads, which have recently invaded the habitat of this species.

Western Pygmy Mulga Snake (*Pseudechis weigeli*), El Questro Station, WA

Genus *Pseudonaja*
Brown Snakes

Pseudonaja comprises nine species of medium-size to large snakes that are widespread throughout Australia; one species also occurs in New Guinea. Three subspecies of *Pseudonaja affinis* previously have been recognized: *P. a. affinis*, the mainland subspecies; *P. a. exilis*, from Rottnest Island; and *P. a. tanneri*, from the islands of the Archipelago of the Recherche. Despite the island races' much smaller size (TL to 1.2 m), their validity as subspecies is unclear, as recent molecular work does not support their distinctiveness.

Brown snakes are usually identified by a single large anterior temporal scale on each side of the head. They are fast-moving, diurnal to nocturnal snakes with moderately large eyes and round pupils. Nervous and defensive, if disturbed they present a cobra-like, hooded threat display. Diet includes lizards, other snakes, and mammals. Most species coil around their prey during capture and ingestion. All *Pseudonaja* species are oviparous. In some areas brown snakes may be confused with mulga snakes (*Pseudechis*) or taipans (*Oxyuranus*).

Pseudonaja venom is neurotoxic, with a strong procoagulant activity, weakly hemolytic, cytotoxic, and myotoxic. Brown snake or polyvalent antivenom is used to neutralize bites from *Pseudonaja* species. However, research has found that the Ringed Brown Snake (*Pseudonaja modesta*) has a radically different venom profile from other species, lacking coagulant activity or neurotoxicity. This is indicative of a potential failure of the brown snake antivenom manufactured by the Commonwealth Serum Laboratories to neutralize envenomation effects from this snake, as the primary antibodies in the antivenom are selected toward the most dominant toxins in the immunizing venom (that of the Eastern Brown Snake, *P. textilis*), and thus would likely be directed predominantly against the dominant toxin type in *P. modesta* venom (Jackson et al 2013). It is still suggested that all *Pseudonaja* species be rated as dangerously venomous at this time.

Dugite
Pseudonaja affinis

TL 2 m. **Lethality.** Dangerously venomous. **Description.** Body large and robust with a slender tail. Coloration very variable, from pale cream to tan, with scattered dark spots and blotches, to almost black. Juvenile and subadult have black head and black band on nape. Head long and narrow, slightly distinct from neck, with straight-edged snout. Ventral area cream to yellow with orange blotches. **Scalation.** Dorsal scales smooth, glossy, and in 19–21 rows at mid-body (eastern populations have 17 rows). Ventrals 190–230. Anal scale divided. Subcaudals 50–70, all divided. **Habitat and range.** Widespread through a variety of habitats, including coastal dunes, heathlands, agricultural lands, and arid shrublands, from sw. WA to the Eyre Peninsula in SA. **Behavior.** Mostly diurnal but nocturnal in warmer weather. A swift snake generally observed basking or foraging. Shelters beneath ground debris and in holes. Diet includes reptiles and small mammals. Clutches of 3–31 eggs recorded. Defensive if disturbed, rearing its forebody in an S shape. **Identification.** *Pseudonaja affinis* overlaps in distribution with *P. aspidorhyncha*, which can be identified by a large, prominent straplike rostral scale; and with *P. inframacula*, which has a dark gray ventral surface. **Conservation.** IUCN status: Least Concern. No threats listed

Dugite (*Pseudonaja affinis*), Esperance, WA. Insert: Dugite (*Pseudonaja affinis*), close-up, Esperance, WA.

Dugite (*Pseudonaja affinis*), Rottnest Island form, Rottnest Island, WA

Dugite (*Pseudonaja affinis*), Archipelago of the Recherche form, Archipelago of the Recherche, WA

Strap-snouted Brown Snake
Pseudonaja aspidorhyncha

TL 1.3 m. **Lethality.** Dangerously venomous. **Description.** Body moderately large and robust with a slender tail. Coloration very variable, from brown to gray-brown or reddish brown, generally with some pattern of black scales on the neck, and in some individuals a series of narrow irregular bands. Juvenile and subadult have black head and black band on nape. Head long and narrow, slightly distinct from neck, with straight-edged snout and large, straplike rostral scale. Ventral area cream, yellow, orange, or gray, with dark gray or dark orange blotches. **Scalation.** Dorsal scales smooth, glossy, and in 17 rows at mid-body. Ventrals 200–230. Anal scale divided. Subcaudals 45–70, all divided. **Habitat and range.** Occurs in mallee woodlands, savanna woodlands, grasslands, and disturbed habitat, in semiarid areas from coastal SA to the arid interior of e. Australia. **Behavior.** Diurnal and nocturnal. A swift species, generally observed basking or foraging. Usually retreats into holes in the ground. Diet includes reptiles and small mammals. Clutches of 9–14 eggs recorded. Defensive if disturbed, rearing its forebody in an S shape. **Identification.** *Pseudonaja aspidorhyncha* is similar to other brown snake species but can be identified by the large, prominent straplike rostral scale and a pale gray mouth lining. May be confused with the Mulga Snake (*Pseudechis australis*), which has mostly single subcaudals; or with the Inland Taipan (*Oxyuranus microlepidotus*), which has 23 mid-body scale rows. **Conservation.** IUCN status: Least Concern. No threats listed.

Strap-snouted Brown Snake (*Pseudonaja aspidorhyncha*), Weengallon, QLD

Strap-snouted Brown Snake (*Pseudonaja aspidorhyncha*), juvenile, Coober Pedy, SA

Strap-snouted Brown Snake (*Pseudonaja aspidorhyncha*), banded form, Cunnamulla, QLD

Strap-snouted Brown Snake (*Pseudonaja aspidorhyncha*), Bollon, QLD

Speckled Brown Snake
Pseudonaja guttata

TL 1.4 m. **Lethality.** Dangerously venomous. **Description.** Body moderately large and robust with a slender tail. Coloration very variable, though usually brown to yellow, cream, or orange; occurs in plain, banded, and speckled forms. Juvenile and subadult have black head and black band on nape. Head long and narrow, slightly distinct from neck, with straight-edged snout. Ventral area yellow with orange-red spots. **Scalation.** Dorsal scales smooth, glossy, and in 19–21 rows at mid-body. Ventrals 190–220. Anal scale divided. Subcaudals 45–70, all divided. **Habitat and range.** Occurs in grassed, black-soil plains of the interior of QLD and adjacent areas of the NT and ne. SA. **Behavior.** Diurnal. A swift species. Shelters in cracking clay soils. Diet includes frogs, reptiles, and small mammals. Clutches of 3–17 eggs recorded. Defensive if disturbed, rearing its forebody in an S shape. **Identification.** *Pseudonaja guttata* is a black-soil specialist and could be confused with *P. ingrami*, which occurs in similar habitat but has a dark tip on each scale. Other similar brown snake species that may overlap in distribution have 17 rows of mid-body scales and are more widely distributed in other habitats. **Conservation.** IUCN status: Least Concern. No threats listed.

Speckled Brown Snake (*Pseudonaja guttata*), plain form, Barkly Tableland, NT

Speckled Brown Snake (*Pseudonaja guttata*), banded form, Barkly Tableland, NT

Speckled Brown Snake (*Pseudonaja guttata*), speckled form, Barkly Tableland, NT

Peninsula Brown Snake
Pseudonaja inframacula

TL 1.6 m. **Lethality.** Dangerously venomous. **Description.** Body large and robust with a slender tail. Coloration variable, from lighter brown to reddish brown or blackish brown, with some darker flecking. Juvenile and subadult have black head and black band on nape. Head long and narrow, slightly distinct from neck, with straight-edged snout. Ventral area immaculate dark gray. **Scalation.** Dorsal scales smooth, glossy, and in 17–19 rows at midbody. Ventrals 190–230. Anal scale divided. Subcaudals 50–65, all divided. **Habitat and range.** Occurs in mallee woodlands, shrublands, coastal heathlands, and disturbed habitats of the Yorke and Eyre Peninsulas of s. SA. An isolated population occurs on Nullarbor Plain, WA. **Behavior.** Diurnal. A swift species generally observed basking or foraging. Usually retreats into vegetation or holes in the ground. Diet includes frogs, reptiles, and small mammals. Clutches of 12 eggs recorded. Defensive if disturbed, rearing its forebody in an S shape. **Identification.** *Pseudonaja inframacula* is distinguished from other brown snake species that overlap in distribution by its immaculate dark gray ventral region. **Conservation.** IUCN status: Least Concern. No threats listed.

Peninsula Brown Snake (*Pseudonaja inframacula*), Nullarbor, SA

Peninsula Brown Snake (*Pseudonaja inframacula*), juvenile, Yorke Peninsula, SA

Ingram's Brown Snake
Pseudonaja ingrami

TL 1.8 m. **Lethality.** Dangerously venomous. **Description.** Body large and robust with a slender tail. Coloration very variable, from reddish brown or dark brown to orange or pale yellow, with most scales dark-tipped. Juvenile and subadult have black head and black band on nape. Head long and narrow, slightly distinct from neck, with straight-edged snout. Mouth lining bluish black. Ventral area yellow with orange-red spots. **Scalation.** Dorsal scales smooth, glossy, and in 19–21 rows at mid-body. Ventrals 190–220. Anal scale divided. Subcaudals 55–70, all divided. **Habitat and range.** Restricted to seasonally flooded grasslands with deep cracking clay soils from the interior of w. QLD to central e. NT; an isolated population occurs near Kununurra, WA. **Behavior.** Diurnal. A swift species generally observed basking or foraging. Shelters beneath ground debris and in deep cracking soil. Diet includes reptiles and small mammals. Clutches of 5–18 eggs recorded. Defensive if disturbed, rearing its forebody in an inverted J shape and flattening the neck region. **Identification.** *Pseudonaja ingrami* is a black-soil specialist and could be confused with *P. guttata*, which occurs in similar habitat but lacks a black tip on each scale. Other similar brown snake species that may overlap in distribution have 17 rows of mid-body scales and are more widely distributed in other habitats. **Conservation.** IUCN status: Least Concern. No threats listed but possibly at some risk from ingesting Cane Toads.

Ingram's Brown Snake (*Pseudonaja ingrami*), Boulia, QLD

Western Brown Snake
Pseudonaja mengdeni

TL 1.2 m. **Lethality.** Dangerously venomous. **Description.** Body moderately large and robust with a slender tail. Coloration very variable, from pale cream to brown, reddish brown, or orange; may be plain, herringbone-patterned, or banded. Juvenile, subadult, and some adults have black head and black band on nape. Head long and narrow, slightly distinct from neck, with rounded snout. Blackish mouth lining. Ventral area cream or yellow with dark red or orange blotches. **Scalation.** Dorsal scales smooth, glossy, and in 17 rows at mid-body. Ventrals 190–220. Anal scale divided. Subcaudals 55–70, all divided. **Habitat and range.** Widespread in a variety of arid habitats, including sand plains, spinifex-dominated dune fields, grasslands, disturbed habitats, stony plains, and ranges, across much of central and w. Australia. **Behavior.** Diurnal and nocturnal. A swift species generally observed basking or foraging. Usually retreats into holes in the ground. Diet includes reptiles and small mammals. Clutches of 7–22 eggs recorded. Defensive if disturbed, rearing its forebody in an S shape. **Identification.** *Pseudonaja mengdeni* is similar to and may overlap in distribution with *P. nuchalis*; both species have a blackish mouth lining, but *P. nuchalis* has a larger rostral scale. *Pseudonaja aspidorhyncha* is also similar but has a large, straplike rostral scale. *Pseudonaja mengdeni* is also sometimes confused with the Inland Taipan (*Oxyuranus microlepidotus*), which has a single anal scale. **Conservation.** IUCN status: Least Concern. No threats listed.

Western Brown Snake (*Pseudonaja mengdeni*), plain form, Alice Springs, NT

Western Brown Snake (*Pseudonaja mengdeni*), black-headed form, Alice Springs, NT

Western Brown Snake (*Pseudonaja mengdeni*), head, Port Hedland, WA

Ringed Brown Snake
Pseudonaja modesta

TL 600 mm. **Lethality.** Venomous, of limited medical significance. **Description.** Body medium-size and slender with a slender tail. Pale gray, tan, or rich reddish brown with a black patch on top of head, a broad black band on nape, and 4–12 widely spaced, narrow dark bands between nape and tail tip; bands may be indistinct or absent on older specimens. Head long and narrow, slightly distinct from neck, with narrowly rounded snout. Ventral area cream flecked with orange. **Scalation.** Dorsal scales smooth and in 17 rows at mid-body. Ventrals 145–175. Anal scale divided. Subcaudals 35–50, all divided. **Habitat and range.** Widespread in most habitats, particularly semiarid shrublands and grasslands, from the interior of QLD and NSW through SA and the NT to WA. **Behavior.** Diurnal and nocturnal. Shelters beneath ground debris and in animal burrows. Average clutches of 6 eggs recorded. Defensive if disturbed, rearing its forebody in an S shape. As a member of the genus *Pseudonaja*, this species should be treated with caution, even though its venom is classified as having limited medical significance. **Identification.** *Pseudonaja modesta* is similar to other brown snake species but can be identified by its smaller size and usual pattern of 4–12 widely spaced, narrow dark bands. It also has a lower ventral scale count, 145–175. **Conservation.** IUCN status: Least Concern. No threats listed.

Ringed Brown Snake (*Pseudonaja modesta*), Mt. Isa, QLD

Northern Brown Snake
Pseudonaja nuchalis

TL 1.3 m. **Lethality.** Dangerously venomous. **Description.** Body moderately large and robust with a slender tail. Coloration very variable, from pale brown to dark brown or golden brown, generally with some pattern of black scales on the neck or a series of bands. Juvenile and subadult have black head and black band on nape. Head long and narrow, slightly distinct from neck, with straight-edged snout and large straplike rostral scale. Ventral area yellow with orange or red spots and blotches. **Scalation.** Dorsal scales smooth, glossy, and in 17 rows at mid-body. Ventrals 180–230. Anal scale divided. Subcaudals 50–70, all divided. **Habitat and range.** Widespread in tropical woodlands, savannas, grasslands, and rocky ranges across n. Australia, from nw. QLD to the e. Kimberley region, WA. **Behavior.** Diurnal and nocturnal. A swift species. Shelters beneath ground debris and in holes in the ground. Diet includes reptiles and small mammals. Clutches of 8–16 eggs recorded. Defensive if disturbed, rearing its forebody in an S shape. **Identification.** *Pseudonaja nuchalis* may overlap in distribution with *P. mengdeni*; both species have a blackish mouth lining, but the latter can be distinguished by a smaller rostral scale. **Conservation.** IUCN status: Data Deficient. Local populations of this species appear to have been reduced since the arrival of Cane Toads in its range.

Northern Brown Snake (*Pseudonaja nuchalis*), Mt. Isa, QLD

Eastern Brown Snake
Pseudonaja textilis

TL 2.2 m. **Lethality.** Dangerously venomous. **Description.** Body large and robust with a slender tail. Coloration very variable, from uniform brown to gray-brown, reddish brown, or almost black; some specimens have banding. Juvenile often has broad black band across head and nape; this pattern generally disappears with maturity. Head long and narrow, slightly distinct from neck, with narrowly rounded snout. Ventral area cream with numerous dark orange blotches. **Scalation.** Dorsal scales smooth, glossy, and in 17 rows at mid-body. Ventrals 180–235. Anal scale divided. Subcaudals 45–75, all divided. **Habitat and range.** Occurs in a wide variety of habitats but mostly associated with dry woodlands, grasslands, rocky plains, and open forests throughout most of e. Australia, extending northward across the NT just into WA, with an apparent isolated population in c. Australia. Also occurs in New Guinea. **Behavior.** Diurnal and nocturnal. A fast-moving, dangerous snake responsible for the greatest number of snakebites in Australia. Diet includes reptiles and small mammals. Average clutches of 16 eggs recorded. Defensive if disturbed, rearing its forebody in an S shape and advancing with mouth open. **Identification.** *Pseudonaja textilis* is similar to other brown snake species but can be identified by the rostral scale not being straplike and a pink mouth lining. It could be confused with the Coastal Taipan (*Oxyuranus scutellatus scutellatus*) in n. Australia, but the latter species has a single anal scale. **Conservation.** IUCN status: Least Concern. No threats listed.

Eastern Brown Snake (*Pseudonaja textilis*), dark form, Canberra, ACT

Eastern Brown Snake (*Pseudonaja textilis*), juvenile, Bandameer, NSW

Eastern Brown Snake (*Pseudonaja textilis*), light form, Mt. Malloy, QLD

Eastern Brown Snake (*Pseudonaja textilis*), Mt. Malloy, QLD

Genus *Tropidechis*
Rough-scaled Snake

Tropidechis is a monotypic genus restricted to e. Australia. The single species is a medium-size snake with keeled scales and moderately large eyes with a round pupil. Prey items include frogs, lizards, birds, and small mammals. This snake is live-bearing.

Venom is strongly coagulant, moderately neurotoxic, hemolytic, and cytotoxic. Tiger Snake or polyvalent antivenom is used to neutralize bites from this species.

Rough-scaled Snake
Tropidechis carinatus

TL 900 mm. **Lethality.** Dangerously venomous. **Description.** Body medium-size and robust with a slender tail. Yellowish brown or olive to dark brown, usually with irregular, narrow dark crossbands, but some individuals plain-patterned. Head long, slightly distinct from neck, with moderately rounded snout. Ventral area cream, yellow, or olive green, often with numerous darker blotches. **Scalation.** Dorsal scales keeled and in 23 rows at mid-body. Ventrals 160–180. Anal scale single. Subcaudals 50–60, all single. **Habitat and range.** Generally encountered around streams and swamps in wet sclerophyll forests and rain forests. Occurs in two separate populations: a southern population from Gosford, NSW, to Fraser Island (K'gari), QLD; and a northern population restricted to the Wet Tropics of ne. QLD. **Behavior.** Nocturnal and diurnal. Mostly terrestrial but readily climbs into low foliage to bask. Litters of 5–19 young recorded. A nervous and very dangerous snake, quickly becoming aggressive if disturbed. **Identification.** Most likely to be confused with the harmless Keelback (*Tropidonophis mairii*), but that species has the anal and subcaudal scales divided and has a loreal scale (present in colubrid snakes, located between the nasal scale and the preocular scale). *Tropidechis carinatus* could also be confused with the Mainland Tiger Snake (*Notechis scutatus scutatus*) in se. QLD, but that species has smooth dorsal scales and only 17 mid-body scale rows. **Conservation.** IUCN status: Least Concern. No threats listed.

Rough-scaled Snake (*Tropidechis carinatus*), Lamington Plateau, QLD

Keelback (Tropidonophis mairii), Enoggera, QLD

Medically Significant Venomous Land Snakes

Eleven species of land snakes categorized by Mirtschin, Rasmussen, and Weinstein (2017) as medically significant: capable of delivering a bite to a human that, if untreated, is likely to have a serious and possibly fatal outcome.

Genus *Cryptophis*
Small-eyed Snakes

This genus contains five species of small to moderate-size snakes occurring in n. and e. Australia. Secretive and terrestrial, they have small eyes, glossy scales, and generally a uniform dorsal coloration. Prey items include lizards and frogs. *Cryptophis* species are live-bearers.

The Eastern Small-eyed Snake (*Cryptophis nigrescens*) is the only species of this genus whose bite is considered to be of medical significance. One fatality has resulted from presumed envenomation, but the death may have been caused by a venom anaphylaxis. The venom contains myotoxins. Tiger snake or polyvalent antivenom is used to neutralize bites from this species. (The other four *Cryptophis* species are classified as potentially dangerous; see pp. 127–131.)

Eastern Small-eyed Snake
Cryptophis nigrescens

TL 500 mm. **Lethality.** Medically significant. **Description.** Body small and robust with a short tapering tail. Uniform black or dark gray. Head relatively depressed with squarish snout; the nasal scale contacts the preocular scale. Ventral area cream to bright pink, usually with black flecks. **Scalation.** Dorsal scales smooth and glossy, in 15 rows at mid-body. Ventrals 165–210. Anal scale single. Subcaudals 30–45, all single. **Habitat and range.** Occurs in rain forests, wet sclerophyll forests, dry sclerophyll forests, and woodlands on the coast and ranges of e. Australia. **Identification.** *Cryptophis nigrescens* is often confused with the Red-belled Black Snake (*Pseudechis porphyriacus*), but the latter species obtains a much larger adult size and has 17 mid-body scale rows. **Behavior.** A secretive, nocturnal snake. Shelters beneath ground debris. Litters of 2–8 young recorded. **Conservation.** IUCN status: Least Concern. No threats listed.

Eastern Small-eyed Snake (*Cryptophis nigrescens*), Cooranbong, NSW

Genus *Demansia*
Whipsnakes

A group of fifteen long, slender species, whipsnakes are found mostly in drier open habitats across Australia. They are active, swift, diurnal snakes with relatively large eyes and round pupils. Their diet consists mostly of lizards. *Demansia* species are oviparous.

The venom contains neurotoxins, and bites are recorded to produce severe local pain and swelling. The two species listed here may deliver bites categorized as medically significant, able to cause life-threatening symptoms. Other species are categorized as potentially dangerous (see pp. 107–194). Tiger Snake or polyvalent antivenom is used to neutralize bites from these species.

Greater Black Whipsnake
Demansia papuensis

TL 1.65 m. **Lethality.** Medically significant. **Description.** Body slender with a long slender tail. Gray to black, tan, or reddish brown and weakly patterned. Head long and tan-colored, sometimes with darker spots, with moderately pointed snout. Creamy-yellow margin around eye. Ventral area gray. **Scalation.** Dorsal scales smooth and in 15 rows at mid-body. Ventrals 198–225. Anal scale divided. Subcaudals 75–110, all divided. **Habitat and range.** Associated mostly with tropical woodlands and grasslands throughout monsoonal n. Australia. **Behavior.** Diurnal. An active, fast-moving species generally observed foraging throughout the day. Shelters beneath ground debris. Diet consists of lizards, frogs, and other snakes. Clutches of 5–13 eggs recorded. **Identification.** *Demansia papuensis* is similar to *D. vestigiata* but has more than 197 ventral scales. Sometimes confused with the Coastal Taipan (*Oxyuranus scutellatus scutellatus*), but that snake has a single anal scale. **Conservation.** IUCN status: Least Concern. No threats listed.

Greater Black Whipsnake (*Demansia papuensis*), Calvert River area, NT

Lesser Black Whipsnake
Demansia vestigiata

TL 1.2 m. **Lethality.** Medically significant. **Description.** Body slender with a long slender tail. Gray to black and sometimes flushed with red posteriorly; individual scales with a darker rear edge form an overall reticulated pattern. Head long, with moderately pointed snout and darker blotches. Creamy-yellow margin around eye. Ventral area gray. **Scalation.** Dorsal scales smooth and in 15 rows at mid-body. Ventrals 165–197. Anal scale divided. Subcaudals 70–95, all divided. **Habitat and range.** Associated mostly with dry forests, woodlands, and grasslands, from se. QLD to Cape York Peninsula and the NT, across the n. Top End to the Kimberley region of WA. Also occurs in New Guinea. **Behavior.** Diurnal. An active, fast-moving species generally observed foraging throughout the day. Shelters beneath ground debris. Diet consists of lizards, frogs, and other snakes. Clutches of 3–17 eggs recorded. **Identification.** *Demansia vestigiata* is similar to *D. papuensis* but has fewer than 197 ventral scales. **Conservation.** IUCN status: Least Concern. No threats listed.

Lesser Black Whipsnake (*Demansia vestigiata*), Mt. Malloy, QLD

Genus *Glyphodon*
Dunmall's and Brown-headed Snakes

The two *Glyphodon* species are moderate-size snakes that occur in e. Australia. They are slender, with glossy, smooth scales, a weakly depressed head, and small black eyes. Prey items consist primarily of skinks. Both species are oviparous.

The venom of these two snakes has been poorly studied, and while the properties of Dunmall's Snake (*Glyphodon dunmalli*) remain unknown, the Brown-headed Snake (*G. tristis*) has been shown to have neurotoxic and myotoxic components. A recorded bite from a Dunmall's Snake produced a severe reaction. No specific antivenom is suggested for the bite of that species, but polyvalent antivenom is recommended for bites of the Brown-headed Snake.

Dunmall's Snake
Glyphodon dunmalli

TL 700 mm. **Lethality.** Medically significant. **Description.** Body medium-size and moderately robust with a short tail. Uniform dark gray or pale brown to blackish brown. Head slightly distinct from neck, with squarish snout and a few pale blotches on upper lip. Ventral area white. **Scalation.** Dorsal scales smooth, glossy, and in 21 rows at mid-body. Ventrals 175–190. Anal scale divided. Subcaudals 35–50, all divided. **Habitat and range.** Associated mostly with woodlands of the Brigalow Belt in se. QLD, north to about Rockhampton and south to adjacent NSW. **Behavior.** Nocturnal. A poorly known terrestrial species that is observed infrequently. Shelters beneath ground debris. Prey items include lizards, frogs, and reptile eggs. Clutches of 5–9 eggs recorded. **Identification.** *Glyphodon dunmalli* can be distinguished from the similar *G. tristis* by the plain nape and 21 mid-body scale rows. The two species also have completely separate geographic ranges. **Conservation.** IUCN status: Data Deficient. The clearing and conversion of native habitat to pastureland represents the biggest threat to this species.

Dunmall's Snake (*Glyphodon dunmalli*), Glenmorgan, QLD

Dunmall's Snake (*Glyphodon dunmalli*), close-up, Glenmorgan, QLD

Brown-headed Snake
Glyphodon tristis

TL 1 m. **Lethality.** Medically significant. **Description.** Body medium-size and moderately robust with a short tail. Dark purplish brown to blackish with reticulated pattern and conspicuous yellow band across nape. Head slightly distinct from neck and paler-colored, with squarish snout. Ventral area white to light gray. **Scalation.** Dorsal scales smooth, glossy, and in 17–19 rows at mid-body. Ventrals 160–190. Anal scale divided. Subcaudals 30–60, all divided. **Habitat and range.** Associated mostly with woodlands, vine thickets, savannas, and monsoon forests of ne. Cape York Peninsula, QLD; Torres Strait Islands; and New Guinea. **Behavior.** Nocturnal. A terrestrial species usually observed moving around at night. Shelters beneath ground debris. Prey items include lizards, frogs, and reptile eggs. Clutches of 6–8 eggs recorded. An extremely nervous species. **Identification.** *Glyphodon tristis* can be distinguished from the similar *G. dunmalli* by the yellow band across the nape and 17–19 mid-body scale rows. The two species also have completely separate distributions. **Conservation.** IUCN status: Least Concern. No threats listed.

Brown-headed Snake (*Glyphodon tristis*), Weipa, QLD

Genus *Hoplocephalus*
Broad-headed Snakes

This genus consists of three species of medium-size to large snakes restricted to woodlands, forests, and rock outcrops in e. Australia. The range of one species includes inland areas, while the other two are associated more with the coast and ranges. They are agile climbers with a noticeably broad, flattened head, distinct from the body. The ventral scales are laterally keeled and notched. Eyes are moderately large with round pupils. Prey items include lizards, frogs, mammals, and birds. *Hoplocephalus* species are live-bearers.

Venom contains procoagulants, myotoxins, and neurotoxins. Bites recorded from all three species have resulted in a rapid onset of severe symptoms, including headaches, swelling, and nausea. A death, possibly from the bite of a Stephens' Banded Snake (*Hoplocephalus stephensii*), was recorded in 2013 in Bellingen, NSW. Tiger Snake or polyvalent antivenom is used to neutralize bites from these species.

Pale-headed Snake
Hoplocephalus bitorquatus

TL 800 mm. **Lethality.** Medically significant. **Description.** Body medium-size and moderately robust with a slender tail. Light brown to gray or black, with head paler than body. Head noticeably distinct from neck, with white to pale gray band across nape, moderately rounded snout, and lips usually barred in dark gray and cream. Ventral area usually creamy gray with some darker flecks. **Scalation.** Dorsal scales smooth and in 19–21 rows at mid-body. Ventrals 190–225. Anal scale single. Subcaudals 40–65, all single. **Habitat and range.** Occurs within dry woodlands, usually along floodplains, from ne. NSW to the Wet Tropics of ne. QLD. **Behavior.** Nocturnal. A terrestrial and arboreal species, generally observed foraging for frogs. Shelters beneath bark and in tree hollows. Diet consists of lizards, frogs, and small mammals. Litters of 2–11 young recorded. Very defensive when disturbed, rearing the forebody into an S shape and delivering rapid strikes. **Identification.** *Hoplocephalus bitorquatus* can be distinguished from the similar *H. stephensii* by the white to pale gray band across its nape. It does not overlap in range with *H. bungaroides.* **Conservation.** IUCN status: Least Concern. Threats include habitat loss and fragmentation.

Pale-headed Snake (*Hoplocephalus bitorquatus*), Moonee, QLD

Broad-headed Snake
Hoplocephalus bungaroides

TL 900 mm. **Lethality.** Medically significant. **Description.** Body medium-size and moderately robust with a slender tail. Black with a very distinctive overall pattern of numerous narrow, irregular bands of bright yellow scales. Head noticeably distinct from neck, with moderately rounded snout. Ventral area gray. **Scalation.** Dorsal scales smooth and in 21 rows at midbody. Ventrals 200–230. Anal scale single. Subcaudals 40–65, all single. **Habitat and range.** Associated with sandstone escarpments and exfoliating rock outcrops within dry woodlands and heaths of the Sydney region and south to Nowra, NSW. **Behavior.** Nocturnal. A terrestrial and arboreal species. Shelters beneath rocks and under loose bark of trees. Diet consists of lizards and occasionally small mammals. Litters of 2–12 young recorded. Very defensive when disturbed, rearing the forebody into an S shape and delivering rapid strikes. **Identification.** *Hoplocephalus bungaroides* can be distinguished from the similar *H. stephensii* by its very distinctive black and yellow coloration. Sometimes confused with harmless juvenile Diamond Python (*Morelia spilota spilota*). **Conservation.** IUCN status: Vulnerable. Threats include habitat loss and fragmentation.

Broad-headed Snake (*Hoplocephalus bungaroides*), Morton National Park, NSW

Diamond Python (*Morelia spilota spilota*), juvenile, Yalwal Plateau, NSW

Stephens' Banded Snake
Hoplocephalus stephensii

TL 1.2 m. **Lethality.** Medically significant. **Description.** Body large and moderately robust with a slender tail. Light brown, yellowish, or dark gray, with series of broad black crossbands, and head paler-colored than body. Some individuals lack banding and are a uniform dark color. Head noticeably distinct from neck, with brown or cream patch on either side of nape, moderately rounded snout, and lips usually with white bars or blotches. Ventral area creamy gray, usually with some darker flecks. **Scalation.** Dorsal scales smooth and in 21 rows at mid-body. Ventrals 225–250. Anal scale single. Subcaudals 50–70, all single. **Habitat and range.** Mostly associated with wet sclerophyll forests and rain forests but also rock outcrops in drier forests. Occurs from Gosford, NSW, to Maryborough, QLD, with isolated populations farther north in Kroombit Tops and Eungella. **Behavior.** Nocturnal. A terrestrial and arboreal species that shelters beneath logs and under the loose bark of trees. Diet consists of lizards, frogs, and small mammals. Litters of 2–17 young recorded. Very defensive when disturbed, rearing the forebody into an S shape and delivering rapid strikes. **Identification.** Usually easily identified by the pale, diffuse cream to brown bands along length of body. Individuals of *Hoplocephalus stephensii* with reduced pattern can be similar to *H. bitorquatus* but may be identified by the brown or cream patch on either side of nape. **Conservation.** IUCN status: Near Threatened. Threats include timber harvesting, forest clearance, and urban development.

Stephens' Banded Snake (*Hoplocephalus stephensii*), banded form, Beechmont, QLD

Stephens' Banded Snake (*Hoplocephalus stephensii*), unbanded form, Tenterfield, NSW

Genus *Paroplocephalus*
Lake Cronin Snake

This genus is represented by a single medium-size, slender species occurring in a small area of sw. WA. Prey items include lizards, frogs, and small mammals. It is live-bearing.

There is one bite record for *Paroplocephalus*. In October 1979, a 24-year-old man was bitten on the finger of his right hand while photographing the holotype of *P. atriceps*. He initially ignored the bite and continued taking photographs, then developed a severe headache as the first noticeable symptom. He started feeling unwell within 15–30 minutes, with generalized diaphoresis and vomiting. He washed the wound and applied a tourniquet to his right arm. He did not have dizziness or syncope but had some pain at the bite site. He arrived at a hospital 40 minutes after the bite. On examination, he was alert but pale and diaphoretic, with a blood pressure level of 170/100 mmHg and a heart rate of 70 beats per minute. There was no evidence of bleeding at mucous membranes or the bite site. He developed venom-induced consumption coagulopathy but no neurotoxicity or myotoxicity, similar to the clinical features of *Hoplocephalus* envenomation. He was treated with polyvalent antivenom and made a full recovery.

Lake Cronin Snake
Paroplocephalus atriceps

TL 570 mm. **Lethality.** Medically significant. **Description.** Body medium-size and slender with a slender tail. Uniform brown to silvery gray with a black head and nape. Head noticeably distinct from neck, with a moderately rounded snout and pale spots on lips. Ventral area gray. **Scalation.** Dorsal scales smooth, matte-textured, and in 17–19 rows at mid-body. Ventrals 175–185. Anal scale single. Subcaudals 45–50, all single. **Habitat and range.** Occurs in dry woodlands and rocky outcrops around Lake Cronin and Peak Eleanora in sw. WA. **Behavior.** Nocturnal. Sometimes arboreal. Shelters beneath rocks. Prey items include lizards, frogs, and small mammals. Litter size not known. Defensive when disturbed. **Identification.** A distinctively colored snake with a very restricted distribution, *Paroplocephalus atriceps* is unlikely to be confused with other species. **Conservation.** IUCN status: Least Concern. No threats listed.

Lake Cronin Snake (*Paroplocephalus atriceps*), Lake Cronin area, WA

Lake Cronin Snake (*Paroplocephalus atriceps*), head, Lake Cronin area, WA

Genus *Suta*
Hooded Snakes

The genus *Suta* contains eleven cryptozoic species distributed mostly throughout arid and semiarid regions of Australia. They are small to medium-size snakes with a broad, moderately depressed head, small eyes, and usually a dark head patch. Prey items are mostly small lizards. *Suta* species are live-bearing.

Two species may inflict bites that are considered medically significant. Venom contains myotoxins and neurotoxins. Bites are recorded to cause headaches, swelling, and nausea, as well as painful axillary lymphadenopathy. Polyvalent antivenom is recommended to neutralize bites from the Curl Snake (*Suta suta*). The death of an adult man was recorded in 2007 after a bite from a third species, the Little Whip Snake (*S. flagellum*), though an anaphylactic reaction was probably involved; for this and other members of the genus, see accounts in Potentially Dangerous Venomous Land Snakes (pp. 107–194).

Little Spotted Snake
Suta punctata

TL 525 mm. **Lethality.** Medically significant. **Description.** Body medium-size and robust with a short tail. Olive brown to rich reddish brown, with a reticulated pattern overall and dark spots and blotches on head, nape, and forebody. Head distinct from neck, moderately depressed, with slightly rounded snout and pale streak on each side of head from snout through eye to temple. Eyes have conspicuous small pupil with pale brown to dark-colored iris. Ventral area white to cream. **Scalation.** Dorsal scales smooth, glossy, and in 15 rows at mid-body. Ventrals 150–215. Anal scale single. Subcaudals 20–40, all single. **Habitat and range.** Mostly associated with spinifex grasslands and arid woodlands from w. QLD through the NT to nw. WA. **Behavior.** Nocturnal. A terrestrial species generally observed moving around at night. Shelters beneath ground debris and vegetation. Prey items include frogs, lizards, and small mammals. Litters of 2–5 young recorded. **Identification.** *Suta punctata* can be distinguished from other members of its genus by the prominent dark spots on the head, neck, and forebody. **Conservation.** IUCN status: Least Concern. No threats listed.

Little Spotted Snake (*Suta punctata*), Camooweal, QLD

Little Spotted Snake (*Suta punctata*), head, Mt. Barnett Station, WA

Curl Snake
Suta suta

TL 600 mm. **Lethality.** Medically significant. **Description.** Body medium-size and robust with a short tail. Pale brown or olive brown to rich reddish brown, with a reticulated pattern that is paler on the lower flanks. Head distinct from neck, moderately depressed, with slightly rounded snout, dark hood, and dark-edged pale stripe on side of head from snout through eye to temple. Eyes have conspicuous small pupil with orange iris. Ventral area white to cream. **Scalation.** Dorsal scales smooth, glossy, and in 19–21 rows at mid-body. Ventrals 150–170. Anal scale single. Subcaudals 20–35, all single. **Habitat and range.** Occurs in a variety of semiarid grasslands, woodlands, open saltbush plains, and heavily cracking clay soils, mostly in arid and semiarid areas of e. Australia. **Behavior.** Nocturnal. A terrestrial species generally observed moving around at night. Shelters beneath ground debris and in deep soil cracks. Prey items include frogs, lizards, and small mammals. Litters of 1–9 young recorded. Adopts a defensive posture if disturbed, curling the body into a tight coil and thrashing around. **Identification.** Associated mostly with harder cracking clay soils, *Suta suta* can be identified by its dark hood and dark-edged pale stripe on side of head. It could be confused with juvenile brown snakes (genus *Pseudonaja*), but those have the anal and subcaudal scales divided. **Conservation.** IUCN status: Least Concern. No threats listed.

Curl Snake (*Suta suta*), Winton, QLD

Potentially Dangerous Venomous Land Snakes

Terrestrial elapid snakes categorized as potentially dangerous should be approached with caution and not assumed to be harmless; although there are no data suggesting that they may be capable of medically significant envenomating, they have not been thoroughly studied (Mirtschin, Rasmussen, and Weinstein 2017).

Based on current evidence or by inference from closely related species that have been studied, there are sixty-five species (one of which is also represented by a subspecies) in sixteen genera that are normally unlikely to deliver a lethal bite to a human but that may produce clinical signs, potentially very serious, in some individuals.

Genus *Antaioserpens*
Burrowing Snakes

This genus contains two species of small, moderately robust snakes that occur in e. Australia. One is associated with dry woodlands, and the other with tropical forests and woodlands. Both have glossy smooth scales in 15 mid-body rows and a weakly shovel-shaped snout. Habits are poorly known. Secretive burrowing snakes, they are usually encountered at night. Diet consists primarily of skinks. Both species are oviparous. These snakes have a small mouth and are disinclined to bite.

Eastern Plain-nosed Burrowing Snake
Antaioserpens albiceps

TL 420 mm. **Lethality.** Potentially dangerous; unlikely to cause significant envenomation. **Description.** Body small and moderately robust with a short tail. Brown to rich orange, with reticulated pattern formed by darker-edged scales and sometimes a dark midline caudal stripe. Head dark gray with pale mottling, a distinctive separated black blotch on neck, weakly shovel-shaped snout, and rostral scale almost as long as broad and not wedge-shaped. Ventral area creamy white. **Scalation.** Dorsal scales smooth, glossy, and in 15 rows at mid-body. Ventrals 135–165. Anal scale divided. Subcaudals 15–25, all divided. **Habitat and range.** Sandy environments within tropical grasslands and woodlands, mostly on Cape York Peninsula and farther south to near Clermont, QLD. **Behavior.** Nocturnal. A poorly known burrowing snake encountered foraging on the surface at night. Clutches of 3 eggs recorded. **Identification.** The two *Antaioserpens* species have completely separate distributions. They are similar to some of the burrowing shovel-nosed snakes (genus *Brachyurophis*), but those species have a wedge-shaped rostral scale and distinctive patterns and coloration. **Conservation.** IUCN status: Least Concern. No threats listed.

Eastern Plain-nosed Burrowing Snake (*Antaioserpens albiceps*), Aurukun, QLD

Warrego Burrowing Snake
Antaioserpens warro

TL 440 mm. **Lethality.** Potentially dangerous; unlikely to cause significant envenomation. **Description.** Body small and moderately robust with a short tail. Pale reddish brown to olive, with irregular darker-edged scales creating a speckled pattern and sometimes a dark midline stripe. Head with blackish hood and separated black collar, weakly shovel-shaped snout, and rostral scale almost as long as broad and not wedge-shaped. Ventral area creamy white. **Scalation.** Dorsal scales smooth, glossy, and in 15 rows at midbody. Ventrals 139–150. Anal scale divided. Subcaudals 15–17, all divided. **Habitat and range.** Recorded from poplar box–*Callitris* pine woodlands along the western edge of the s. Brigalow Belt in the n. Murray-Darling Basin in QLD. **Behavior.** Nocturnal. A poorly known burrowing snake encountered foraging on the surface at night. **Identification.** The two *Antaioserpens* species have completely separate distributions. They are similar to some of the burrowing shovel-nosed snakes (genus *Brachyurophis*), but those species have a wedge-shaped rostral scale and distinctive patterns and coloration. **Conservation.** IUCN status: Data Deficient. Threatened by the processes of habitat degradation through cattle grazing, invasive grasses, and altered fire regimes. The species' tolerance of habitat modification is unknown.

Warrego Burrowing Snake (*Antaioserpens warro*), Chesterton Range National Park, QLD

Genus *Brachyurophis*
Shovel-nosed Snakes

These eight species (one represented by two subspecies) of small, moderately robust snakes are widely distributed throughout Australia and associated mostly with drier, warm areas in woodlands, grasslands, and deserts. Most species are colorful and patterned with a series of broad or ragged-edged bands. All have smooth, glossy scales in 15–17 mid-body rows and a shovel-shaped snout. These are secretive burrowing snakes usually encountered at night. Diet consists primarily of reptile eggs and occasionally small skinks. They are oviparous.

The shovel-nosed snakes have a small mouth and are disinclined to bite; however, they possess functional fangs. A bite from one species, the Northern Shovel-nosed Snake (*Brachyurophis roperi*), resulted in localized muscle pain and inflammation radiating up the entire affected arm for several hours.

Northwestern Shovel-nosed Snake
Brachyurophis approximans

TL 360 mm. **Lethality.** Potentially dangerous; unlikely to cause significant envenomation. **Description.** Body small and moderately robust with a short tail. Dark brown to charcoal with cream to gray narrow, ragged-edged bands, a broad black band on head, and another on nape. Head with projecting shovel-shaped snout and wedge-shaped rostral scale. Ventral area cream-colored. **Scalation.** Dorsal scales smooth, glossy, and in 17 rows at mid-body. Ventrals 151–181. Anal scale divided. Subcaudals 19–27, divided. **Habitat and range.** Occurs in acacia woodlands and shrublands of the upper w. coast and arid interior of WA. **Behavior.** Nocturnal. A burrowing snake usually encountered moving around on the surface only at night. Shelters beneath ground debris. Clutches of 2–4 eggs recorded. **Identification.** *Brachyurophis approximans* may overlap in distribution with other *Brachyurophis* species, but each can be distinguished by a different pattern and coloration. **Conservation.** IUCN status: Least Concern. No known major threats.

Northwestern Shovel-nosed Snake (*Brachyurophis approximans*), Indee, WA

Australian Coral Snake
Brachyurophis australis

TL 340 mm. **Lethality.** Potentially dangerous; unlikely to cause significant envenomation. **Description.** Body small and moderately robust with a short tail. Pink, pale reddish brown, or red with narrow, ragged-edged bands formed by pale scales with dark edges. Broad black band on head and another on nape. Head with projecting shovel-shaped snout and wedge-shaped rostral scale. Ventral area cream-colored. **Scalation.** Dorsal scales smooth, glossy, and in 17 rows at mid-body. Ventrals 140–170. Anal scale divided. Subcaudals 15–30, divided. **Habitat and range.** Occurs in a variety of habitats from rock outcrops in dry sclerophyll forests to mallee woodlands, mostly throughout the interior of e. Australia but also in coastal areas of se. and ne.

QLD. **Behavior.** Nocturnal. A burrowing snake usually encountered on the surface only at night. Shelters beneath rocks, logs, and other ground debris. Clutches of 4–6 eggs recorded. **Identification.** *Brachyurophis australis* may overlap in distribution with other *Brachyurophis* species, but each can be distinguished by a different pattern and coloration. **Conservation.** IUCN status: Least Concern. Threats may include predation by feral animals and habitat modification for agriculture.

Australian Coral Snake (*Brachyurophis australis*), Gluepot Reserve, SA

Cape York Shovel-nosed Snake
Brachyurophis campbelli

TL 350 mm. **Lethality.** Potentially dangerous; unlikely to cause significant envenomation. **Description.** Body small and moderately robust with a short tail. Pale gray to pink, orange, or reddish brown, with broad and narrow darker bands, broad black blotch on head, and broad black band on nape. Head with projecting shovel-shaped snout and wedge-shaped rostral scale. Ventral area cream-colored. **Scalation.** Dorsal scales smooth, glossy, and in 15–17 rows at mid-body. Ventrals 140–190. Anal scale divided. Subcaudals 14–30, divided. **Habitat and range.** Found in tropical, seasonally dry woodlands, mostly on Cape York Peninsula and as far south as Longreach, QLD. **Behavior.** Nocturnal. A burrowing snake usually encountered on the surface only at night. Shelters beneath ground debris. Clutches of 6 eggs recorded. **Identification.** *Brachyurophis campbelli* may overlap in distribution with other *Brachyurophis* species, but each can be distinguished by a different pattern and coloration. **Conservation.** IUCN status: Least Concern. No known major threats.

Cape York Shovel-nosed Snake (*Brachyurophis campbelli*), Winton, QLD

Western Narrow-banded Shovel-nosed Snake
Brachyurophis fasciolatus fasciolatus

TL 390 mm. **Lethality.** Potentially dangerous; unlikely to cause significant envenomation. **Description.** Body small and moderately robust with a short tail. White to cream with pale pink and orange flecks; narrow, ragged-edged bands formed by pale scales with black or dark brown edges; and a broad black band on head and another on nape. Head with projecting shovel-shaped snout and wedge-shaped rostral scale. Ventral area cream-colored. **Scalation.** Dorsal scales smooth, glossy, and in 17 rows at mid-body. Ventrals 140–175. Anal scale divided. Subcaudals 15–30, divided. **Habitat and range.** Associated with sandy soils in arid shrublands, spinifex deserts, and coastal dunes from Perth east to Laverton, in the Goldfields area, and north to Shark Bay, WA. **Behavior.** Nocturnal. A burrowing snake usually encountered on the surface only at night. Clutches of 2–5 eggs recorded. **Identification.** *Brachyurophis fasciolatus fasciolatus* can be distinguished from *B. f. fasciatus* by its broader dark bands on head and nape and restricted distribution. Its pattern distinguishes it from other *Brachyurophis* species. **Conservation.** IUCN status: Least Concern. No major threats but unlikely to persist in areas with compacted soils such as agricultural land.

Western Narrow-banded Shovel-nosed Snake (*Brachyurophis fasciolatus fasciolatus*), Jurien Bay, WA

Eastern Narrow-banded Shovel-nosed Snake
Brachyurophis fasciolatus fasciatus

TL 390 mm. **Lethality.** Potentially dangerous; unlikely to cause signifi-cant envenomation. **Description.** Body small and moderately robust with a short tail. Pale reddish to cream with pale pink to orange flecks; narrow, ragged-edged bands formed by pale scales with black or dark brown edges; and a broad black band on head and another on nape. Head with projecting shovel-shaped snout and wedge-shaped rostral scale. Ventral area cream-col-ored. **Scalation.** Dorsal scales smooth, glossy, and in 17 rows at mid-body. Ventrals 140–171. Anal scale divided. Subcaudals 19–27, divided. **Habitat and range.** Associated with sandy soils in arid shrublands, spinifex deserts, and coastal areas from e. WA through most of SA and the s. NT to sw. QLD and nw. NSW. **Behavior.** Nocturnal. A burrowing snake usually encountered on the surface only at night. Clutches of 4–7 eggs recorded. **Identification.** *Brachyurophis fasciolatus fasciatus* can be distinguished from *B. f. fasciolatus* by its narrower dark bands on head and nape and widespread distribution. Its pattern dis-tinguishes it from other *Brachyurophis* species. **Conservation.** IUCN status: Least Concern. Unlikely to be threatened, as most of range is sparsely populated, with little human activity.

Eastern Narrow-banded Shovel-nosed Snake (*Brachyurophis fasciolatus fasciatus*), Ilkulka, WA

Unbanded Shovel-nosed Snake
Brachyurophis incinctus

TL 360 mm. **Lethality.** Potentially dangerous; unlikely to cause significant envenomation. **Description** .Body small and moderately robust with a short tail. Body and tail uniform pale brown to dark pink, without bands, sometimes with dark-edged individual scales forming obscure reticulated pattern. Broad black band on head and another on nape. Head with projecting shovel-shaped snout and wedge-shaped rostral scale. Ventral area cream-colored. **Scalation.** Dorsal scales smooth, glossy, and in 17 rows at mid-body. Ventrals 140–165. Anal scale divided. Subcaudals 18–30, divided. **Habitat and range.** Occurs in a variety of habitats, including stony hills and rock outcrops with woodlands to shrublands and spinifex grasslands, throughout the interior of QLD and the NT. **Behavior.** Nocturnal. A burrowing snake usually encountered on the surface only at night. Clutches of 3–5 eggs recorded. **Identification.** *Brachyurophis incinctus* can be distinguished from other *Brachyurophis* species by its distinctive coloration. **Conservation.** IUCN status: Least Concern. No known major threats.

Unbanded Shovel-nosed Snake (*Brachyurophis incinctus*), Mt. Isa, QLD

Arnhem Shovel-nosed Snake
Brachyurophis morrisi

TL 300 mm. **Lethality.** Potentially dangerous; unlikely to cause significant envenomation. **Description.** Body small and moderately robust with a short tail. Body and tail orange-brown, without bands or dark-edged individual scales forming obscure reticulated pattern. Prominent black blotch on nape. Head pale yellow with reduced black blotch, projecting shovel-shaped snout, and wedge-shaped rostral scale. Ventral area cream-colored. **Scalation.** Dorsal scales smooth, glossy, and in 15 rows at mid-body. Ventrals 135–145. Anal scale divided. Subcaudals 15–25, divided. **Habitat and range.** Occurs in sandy soils throughout monsoonal woodlands of n. Arnhem Land in the NT. **Behavior.** Nocturnal. A burrowing snake usually encountered on the surface only at night. Shelters beneath rocks, logs, and other ground debris. Considered to be oviparous. **Identification.** *Brachyurophis morrisi* may overlap in distribution with other *Brachyurophis* species, but each can be distinguished by a different pattern and coloration. **Conservation.** IUCN status: Least Concern. No known major threats.

Arnhem Shovel-nosed Snake (*Brachyurophis morrisi*), Cobourg Peninsula, NT

Northern Shovel-nosed Snake
Brachyurophis roperi

TL 370 mm. **Lethality.** Potentially dangerous; unlikely to cause significant envenomation. **Description.** Body small and moderately robust with a short tail. Dark brown to orange or reddish brown, with broad, ragged-edged darker bands and broad black band on head and another on nape. Head with projecting shovel-shaped snout and wedge-shaped rostral scale. Ventral area cream-colored. **Scalation.** Dorsal scales smooth, glossy, and in 15–17 rows at mid-body. Ventrals 150–180. Anal scale divided. Subcaudals 15–24, divided. **Habitat and range.** Occurs on heavy clay to rocky soils in tropical, seasonally dry woodlands in the NT from the Kimberley through the Top End and south to about Ti Tree. **Behavior.** Nocturnal. A burrowing snake usually encountered on the surface only at night. Shelters beneath rocks, logs, and other ground debris. Clutches of up to 5 eggs recorded. **Identification.** *Brachyurophis roperi* may overlap in distribution with other *Brachyurophis* species, but each can be distinguished by a different pattern and coloration. **Conservation.** IUCN status: Least Concern. No known major threats.

Northern Shovel-nosed Snake (*Brachyurophis roperi*), Adelaide River, NT

Northern Shovel-nosed Snake (*Brachyurophis roperi*), juvenile, Robinson River area, NT

Southern Shovel-nosed Snake
Brachyurophis semifasciatus

TL 355 mm. **Lethality.** Potentially dangerous; unlikely to cause significant envenomation. **Description.** Body small and moderately robust with a short tail. Orange to reddish brown, with numerous darker bands about the same width as pale interspaces, and a broad black band on head and another on nape. Head with projecting shovel-shaped snout and wedge-shaped rostral scale. Ventral area cream-colored. **Scalation.** Dorsal scales smooth, glossy, and in 15–17 rows at mid-body. Ventrals 150–180. Anal scale divided. Subcaudals 15–24, divided. **Habitat and range.** Occurs in sandy and stony soils in arid scrublands and hummock grasslands to coastal dunes and heaths in s. WA, excluding the far south and southeast, to nw. SA and neighboring border region of the NT. **Behavior.** Nocturnal. A burrowing snake usually encountered on the surface only at night. Shelters beneath rocks, logs, and other ground debris. Clutches of up to 5 eggs recorded. **Identification.** *Brachyurophis semifasciatus* may overlap in distribution with other *Brachyurophis* species, but each can be distinguished by a different pattern and coloration. **Conservation.** IUCN status: Least Concern. No known major threats.

Southern Shovel-nosed Snake (*Brachyurophis semifasciatus*), Yulara area, NT

Genus *Cacophis*
Crowned Snakes

Four species of small to medium-size, robust snakes that occur along the coast and ranges of e. Australia, crowned snakes are associated mostly with rain forests or wet sclerophyll forests. They have glossy, smooth scales in 15 mid-body rows and a pale band or collar on the nape. Secretive, nocturnal snakes, they shelter beneath ground debris. Diet consists primarily of skinks. They are oviparous. If disturbed, crowned snakes will raise the forebody and lunge at the aggressor, though they are disinclined to bite.

Northern Dwarf Crowned Snake
Cacophis churchilli

TL 450 mm. **Lethality.** Potentially dangerous. **Description.** Body small and robust with a short tapering tail. Dark blue-gray to brown, with a narrow, pale yellow band across nape, and darker areas continuing on head and snout. Head not distinct from neck, with narrowly rounded snout. Ventral area gray. **Scalation.** Dorsal scales smooth, glossy, and in 15 rows at mid-body. Ventrals 160–175. Anal scale divided. Subcaudals 20–30, all divided. **Habitat and range.** Occurs in wet sclerophyll forests and rain forests along the coast and ranges of ne. QLD from Mossman to Paluma. **Behavior.** Nocturnal. Clutches of 7–9 eggs recorded. **Identification.** *Cacophis churchilli* is the only species of crowned snake in its restricted area of distribution. **Conservation.** IUCN status: Least Concern. No known major threats.

Northern Dwarf Crowned Snake (*Cacophis churchilli*), Magnetic Island, QLD

White-crowned Snake
Cacophis harriettae

TL 500 mm. **Lethality.** Potentially dangerous. **Description.** Body small and robust with a short tapering tail. Gray to dark brown or almost black, with broad white or cream band across nape, and darker areas continuing along side of head and snout. Head not distinct from neck, with narrowly rounded snout. Ventral area uniform gray. **Scalation.** Dorsal scales smooth, glossy, and in 15 rows at mid-body. Ventrals 170–200. Anal scale divided. Subcaudals 25–45, all divided. **Habitat and range.** Occurs in wet sclerophyll forests and rain forests in se. QLD and ne. NSW. **Behavior.** Nocturnal and secretive. Often shelters under logs or deep leaf litter. Clutches of 2–10 eggs recorded. **Identification.** *Cacophis harriettae* can be distinguished from the similar *C. krefftii* and *C. squamulosus* by the broad white or cream band across the nape, at least five scales wide. **Conservation.** IUCN status: Least Concern. No substantial threats recognized.

White-crowned Snake (*Cacophis harriettae*), Gladstone, QLD

Southern Dwarf Crowned Snake
Cacophis krefftii

TL 345 mm. **Lethality.** Potentially dangerous. **Description.** Body small and robust with a short tapering tail. Dark blue-gray to dark brown or almost black, with narrow, pale yellow band across nape, and darker areas continuing on head and snout. Head not distinct from neck, with narrowly rounded snout. Ventral area pale yellow with narrow dark bands. **Scalation.** Dorsal scales smooth, glossy, and in 15 rows at mid-body. Ventrals 140–160. Anal scale divided. Subcaudals 25–40, all divided. **Habitat and range.** Occurs in wet sclerophyll forests and rain forests along the coast and ranges of se. QLD and ne. NSW. **Behavior.** Nocturnal and secretive. Shelters in moist areas under logs and other ground debris. Clutches of 2–5 eggs recorded. **Identification.** *Cacophis krefftii* can be distinguished from the similar *C. harriettae* and *C. squamulosus* by the narrow pale yellow band across the nape, at most three scales wide. **Conservation.** IUCN status: Least Concern. Occurs in many protected areas. No known major threats.

Southern Dwarf Crowned Snake (*Cacophis krefftii*), Newcastle, NSW

Golden-crowned Snake
Cacophis squamulosus

TL 750 mm. **Lethality.** Potentially dangerous. **Description.** Body medium-size and robust with a short tapering tail. Dark gray to dark brown or almost black, with pale brown to yellow streak along each side of neck, not meeting to form nape band, and darker areas continuing along side of head and snout. Head not distinct from neck, with narrowly rounded snout. Ventral area pink to orange with midline of darker spots or blotches. **Scalation.** Dorsal scales smooth, glossy, and in 15 rows at mid-body. Ventrals 165–185. Anal scale divided. Subcaudals 30–50, all divided. **Habitat and range.** Occurs along the coast and ranges in wet sclerophyll forests and rain forests from Wollongong, NSW, to central e. QLD. Also recorded from dry sclerophyll forests in the ne. Murray-Darling Basin, QLD. **Behavior.** Nocturnal and secretive. Shelters in moist areas under logs and other ground debris. Clutches of 2–15 eggs recorded. **Identification.** *Cacophis squamulosus* can be distinguished from the similar *C. harriettae* and *C. krefftii* by the pale brown to yellow streak along each side of neck, not meeting to form a band. **Conservation.** IUCN status: Least Concern. Known to occur in many protected areas, but no threats listed.

Golden-crowned Snake (*Cacophis squamulosus*), Mt. Glorious, QLD

Genus *Cryptophis*
Small-eyed Snakes

See complete genus description in the Medically Significant Venomous Land Snakes section (p. 85). The four species listed here are potentially dangerous. All species should be treated with caution.

Carpentaria Snake
Cryptophis boschmai

TL 560 mm. **Lethality.** Potentially dangerous. **Description.** Body medi-um-size and robust with a short tapering tail. Dark brown to orange-brown with paler lateral scales, and sides of head yellowish. Head relatively de-pressed, with squarish snout. Ventral area creamy white. **Scalation.** Dorsal scales smooth, glossy, and in 15 rows at mid-body. Ventrals 145–190. Anal scale single. Subcaudals 20–35, all single. **Habitat and range.** Occurs mainly in dry sclerophyll forests and woodlands of e. QLD to w. Cape York Peninsula. **Behavior.** Nocturnal and secretive. Shelters beneath ground de-bris. Litters of 5–11 young recorded. **Identification.** *Cryptophis boschmai* is similar to *C. nigrescens* (p. 86) but can be distinguished by its brown color and the nasal scale not being in contact with the preocular scale. It is distinguished from *C. nigrostriatus* by lack of dark vertebral stripe. **Conservation.** IUCN status: Least Concern. Occurs in several protected areas. No known major threats.

Carpentaria Snake (*Cryptophis boschmai*), Townsville, QLD

Pink Snake
Cryptophis incredibilis

TL 400 mm. **Lethality.** Potentially dangerous. **Description.** Body small and slender with a long tapering tail. Bright pink dorsum. Head relatively depressed, with squarish snout. Ventral area pearly white. **Scalation.** Dorsal scales smooth, glossy, and in 15 rows at mid-body. Ventrals 180–185. Anal scale single. Subcaudals 50–65, all single. **Habitat and range.** Occurs in open eucalypt and paperbark woodlands on Prince of Wales Island in s. Torres Strait, QLD. May have a wider distribution. **Behavior.** A secretive snake known from only a few specimens. Ecology largely unknown. Shelters beneath ground debris on sandy soils. **Identification.** With its bright pink coloration, *Cryptophis incredibilis* is unlikely to be confused with any other species. It is possibly a color variant of *C. nigrostriatus.* **Conservation.** IUCN status: Least Concern. No known major threats.

Pink Snake (*Cryptophis incredibilis*), Prince of Wales Island, QLD

Black-striped Snake
Cryptophis nigrostriatus

TL 500 mm. **Lethality.** Potentially dangerous. **Description.** Body small and slender with a long tapering tail. Dark brown to pinkish red with a broad, dark vertebral stripe from nape to tail tip. Head relatively depressed, with squarish snout. Ventral area pearly white. **Scalation.** Dorsal scales smooth, glossy, and in 15 rows at mid-body. Ventrals 160–190. Anal scale single. Subcaudals 45–75, all single. **Habitat and range.** Occurs in rain forests and woodlands of e. QLD from Rockhampton through Cape York Peninsula to islands of Torres Strait. **Behavior.** Nocturnal and secretive. Shelters beneath ground debris. Litters of 4–9 young recorded. **Identification.** *Cryptophis nigrostriatus* can be distinguished from the similar *C. nigrescens* and *C. boschmai* by its dark vertebral stripe. **Conservation.** IUCN status: Least Concern. No known major threats.

Black-striped Snake (*Cryptophis nigrostriatus*), Laura, QLD

Northern Small-eyed Snake
Cryptophis pallidiceps

TL 550 mm. **Lethality.** Potentially dangerous. **Description.** Body medium-size and robust with a short tapering tail. Glossy black to dark gray or brown, paler on head, with orange tinge on lower flanks. Head relatively depressed, with squarish snout. Ventral area pearly white to pinkish. **Scalation.** Dorsal scales smooth, glossy, and in 15 rows at mid-body. Ventrals 160–180. Anal scale single. Subcaudals 35–55, all single. **Habitat and range.** Occurs in sedgelands, forests, and woodlands of the Top End of the NT and the n. Kimberley of n. WA. **Behavior.** Nocturnal and secretive. Shelters beneath ground debris. Litters of 2–5 young recorded. **Identification.** *Cryptophis pallidiceps* is the only member of this genus that occurs within its area of distribution. **Conservation.** IUCN status: Least Concern. No known major threats.

Northern Small-eyed Snake (*Cryptophis pallidiceps*), Doongan Station, WA

Genus *Demansia*
Whipsnakes

See complete genus description in the Medically Significant Venomous Land Snakes section (p. 84). The venom of *Demansia* species contains neurotoxins, and bites are recorded to produce severe local pain and swelling. The thirteen snakes listed here are regarded as potentially dangerous.

Narrow-headed Whipsnake
Demansia angusticeps

TL 880 mm. **Lethality.** Potentially dangerous. **Description.** Body slender with a long slender tail. Olive brown to gray with the anterior edge of each mid-body scale marked with black. Head long, with moderately pointed snout, distinctive dark comma-shaped mark from eye to corner of mouth, creamy-yellow margin around eye, prominent pale bar on front and rear edges of eye, and pale-edged dark line across front of snout. Ventral area cream to yellowish, often with gray flecks anteriorly. **Scalation.** Dorsal scales smooth, matte-textured, and in 15 rows at mid-body. Ventrals 180–200. Anal scale divided. Subcaudals 70–100, all divided. **Habitat and range.** Occurs in rocky woodlands and tropical savannas from Broome through the Kimberley in n. WA to the Victoria River region of the nw. NT. **Behavior.** Diurnal. An active, fast-moving species generally observed foraging throughout the day. Clutch size not known. **Identification.** *Demansia angusticeps* may be confused with other *Demansia* species within its range but can be distinguished by the narrow pale-edged dark line across front of snout and prominent pale bars on front and rear edges of eye. **Conservation.** IUCN status: Least Concern. No known major threats.

Narrow-headed Whipsnake (*Demansia angusticeps*), Halls Creek, WA

Black-necked Whipsnake
Demansia calodera

TL 610 mm. **Lethality.** Potentially dangerous. **Description.** Body slender with a long slender tail. Olive brown to gray, often with dark spots on dorsal scales forming a reticulated pattern. Head long, with moderately pointed snout, distinctive dark comma-shaped mark from eye to corner of mouth, dark brown margin around eye, pale-edged dark line across snout, and broad pale-edged dark band across neck. Ventral area white, cream, or pale yellow. **Scalation.** Dorsal scales smooth, matte-textured, and in 15 rows at mid-body. Ventrals 170–195. Anal scale divided. Subcaudals 65–90, all divided. **Habitat and range.** Occurs in arid grasslands and shrublands on the central w. coast of WA, including adjacent islands. Also, an apparently isolated population occurs in the interior of the Gibson Desert Nature Reserve in central e. WA. **Behavior.** Diurnal. An active, fast-moving species generally observed foraging throughout the day. Clutch size not known. **Identification.** *Demansia calodera* may overlap with *D. reticulata* and *D. rufescens*, the only other whipsnakes to occur within its range. Those species lack a dark band across the neck. **Conservation.** IUCN status: Least Concern. No known major threats.

Black-necked Whipsnake (*Demansia calodera*), Tamala Homestead, WA

Long-tailed Whipsnake
Demansia flagellatio

TL 715 mm. **Lethality.** Potentially dangerous. **Description.** Body slender with a long slender tail. Reddish brown to bluish gray. Head long and dark, with two prominent yellow or orange bands, dark comma-shaped mark from eye to corner of mouth, dark line between nostrils, and moderately pointed snout. Ventral area white, cream, or yellow. **Scalation.** Dorsal scales smooth, matte-textured, and in 15 rows at mid-body. Ventrals 195–215. Anal scale divided. Subcaudals 65–90, all divided. **Habitat and range.** Occurs in tropical savannas and spinifex-dominated rocky woodlands in nw. QLD from Mt. Isa to Boodjamulla (Lawn Hill) National Park. **Behavior.** Diurnal. An active, fast-moving species generally observed foraging throughout the day. Clutch size not known. **Identification.** *Demansia flagellatio* can be distinguished from other *Demansia* species within its range by its dark head with prominent yellow or orange bands. **Conservation.** IUCN status: Least Concern. No known major threats.

Long-tailed Whipsnake (*Demansia flagellatio*), Boodjamulla (Lawn Hill) National Park, QLD

Olive Whipsnake
Demansia olivacea

TL 835 mm. **Lethality.** Potentially dangerous. **Description.** Body slender with a long slender tail. Olive brown to gray, often with anterior edge of each mid-body scale marked with black. Head long, with moderately pointed snout, weak dark comma-shaped mark from eye to corner of mouth, and dark variegations on lips and snout. Ventral area yellow to bluish yellow. **Scalation.** Dorsal scales smooth, matte-textured, and in 15 rows at mid-body. Ventrals 160–210. Anal scale divided. Subcaudals 65–110, all divided. **Habitat and range.** Occurs in rocky woodlands and savannas of the n. NT and the Kimberley region of WA. **Behavior.** Diurnal. An active, fast-moving species generally observed foraging throughout the day. Clutches of 3–5 eggs recorded. **Identification.** *Demansia olivacea* can be distinguished from other *Demansia* species within its range by having little or no pale-edged dark line across snout, no pale bars on front and rear edges of eyes, and the dark variegations on its lips and snout. **Conservation.** IUCN status: Least Concern. No known major threats.

Olive Whipsnake (*Demansia olivacea*), Doongan Station, WA

Yellow-faced Whipsnake
Demansia psammophis

TL 1 m. **Lethality.** Potentially dangerous. **Description.** Body slender with a long slender tail. Olive brown to gray, often with reddish brown on vertebral region of neck and anterior part of body. Head long and light-colored, with moderately pointed snout, distinctive dark comma-shaped mark from eye to corner of mouth, creamy-yellow margin around eye, and pale-edged dark band across snout. Ventral area gray-green to yellowish. **Scalation.** Dorsal scales smooth and in 15 rows at mid-body. Ventrals 165–230. Anal scale divided. Subcaudals 60–105, all divided. **Habitat and range.** Occurs in a variety of habitats throughout much of e. Australia; associated mostly with dry woodlands and open forests but also occurs in mallee woodlands with spinifex. **Behavior.** Diurnal. An active, fast-moving species generally observed foraging throughout the day. Clutches of 5 or 6 eggs recorded. **Identification.** *Demansia psammophis* is similar to other *Demansia* species within its range and can be distinguished by a light-colored head lacking dark blotches or bar across nape. **Conservation.** IUCN status: Least Concern. No known major threats.

Yellow-faced Whipsnake (*Demansia psammophis*), The Gap, QLD

Somber Whipsnake
Demansia quaesitor

TL 735 mm. **Lethality.** Potentially dangerous. **Description.** Body slender with a long slender tail. Brown to gray, often with yellow on posterior part of body and tail. Variable in coloration across range, with eastern and western populations significantly different, including rust-colored head in east and dark gray head in west. Head long, with moderately pointed snout, dark comma-shaped mark from eye to corner of mouth, dark line across front of snout, and faint darker band across nape. Ventral area pale and patternless, sometimes with dark flecks on throat. **Scalation.** Dorsal scales smooth, matte-textured, and in 15 rows at mid-body. Ventrals 180–200. Anal scale divided. Subcaudals 60–100, all divided. **Habitat and range.** Occurs in tropical woodlands, rocky savannas, and scrubland habitats from the Kimberley region of n. WA through the NT to sw. QLD. **Behavior.** Diurnal. An active, fast-moving species generally observed foraging throughout the day. Clutches of 6 eggs recorded. **Identification.** *Demansia quaesitor* is similar to other *Demansia* species within its range and is distinguished by a dark band across nape lacking a pale edge. **Conservation.** IUCN status: Least Concern. No known major threats.

Somber Whipsnake (*Demansia quaesitor*), Mary Kathleen Dam, QLD

Somber Whipsnake (*Demansia quaesitor*)

Reticulated Whipsnake
Demansia reticulata

TL 1 m. **Lethality.** Potentially dangerous. **Description.** Body slender with a long slender tail. Olive green to gray or yellow, with each scale broadly dark-edged, forming a reticulated pattern. Head long, with moderately pointed snout, distinctive dark comma-shaped mark from eye to corner of mouth, dark brown margin around eye, and pale-edged dark band across snout. Ventral area white to yellow. **Scalation.** Dorsal scales smooth and in 15 rows at mid-body. Ventrals 165–217. Anal scale divided. Subcaudals 70–102, all divided. **Habitat and range.** Occurs in a variety of habitats including coastal heaths, deserts, woodlands, and scrublands from the w. coast of Australia eastward to the arid interior of SA and the NT. **Behavior.** Diurnal. An active, fast-moving species generally observed foraging throughout the day. Shelters beneath ground debris and vegetation. Clutches of up to 6 eggs recorded. **Identification.** *Demansia reticulata* can be distinguished from other *Demansia* species within its range by a plain-colored head and noticeably reticulated pattern overall. **Conservation.** IUCN status: Least Concern. No known major threats.

Reticulated Whipsnake (*Demansia reticulata*), Jurien Bay, WA

Centralian Whipsnake
Demansia cyanochasma

TL 1 m. **Lethality.** Potentially dangerous. **Description.** Body slender with a long slender tail. Olive to gray, with each scale broadly dark-edged, forming a reticulated pattern, and head, forebody, and tail flushed orange to copper. Head long, with moderately pointed snout, distinctive dark comma-shaped mark from eye to corner of mouth, creamy-yellow margin around eye, and pale-edged dark band across snout. Ventral area white to yellow. **Scalation.** Dorsal scales smooth and in 15 rows at mid-body. Ventrals 175–200. Anal scale divided. Subcaudals 60–105, all divided. **Habitat and range.** Occurs in a variety of habitats including arid woodlands and shrublands from the e. Goldfields of WA to se. SA. Also recorded in sw. QLD. **Behavior.** Diurnal. An active, fast-moving species generally observed foraging throughout the day. Shelters beneath ground debris and vegetation. Clutches of 5–8 eggs recorded. **Identification.** *Demansia cyanochasma* can be distinguished from other *Demansia* species within its range by its noticeable reticulated pattern and the orange to copper flush on its head, forebody, and tail. **Conservation.** IUCN status: Least Concern. No known major threats.

Centralian Whipsnake (*Demansia cyanochasma*), Eyre Peninsula, SA

Blacksoil Whipsnake
Demansia rimicola

TL 975 mm. **Lethality.** Potentially dangerous. **Description.** Body slender with a long slender tail. Gray to olive or yellow-brown with individual scale colors forming dark and light lateral stripes. Head long, with moderately pointed snout, distinctive dark comma-shaped mark from eye to corner of mouth, creamy-yellow margin around eye, pale-edged dark band across snout, and pale-edged dark band across nape. Ventral area yellowish to bright orange-red with paired dark brown spots under throat. **Scalation.** Dorsal scales smooth and in 15 rows at mid-body. Ventrals 175–205. Anal scale divided. Subcaudals 65–100, all divided. **Habitat and range.** Inhabits cracking clay soils in drier open grasslands and shrublands through inland areas of nw. NSW, ne. SA, QLD, and the NT to nw. WA. **Behavior.** Diurnal. An active, fast-moving species generally observed briefly before it retreats into vegetation or cracks in the ground. Clutch size not known. **Identification.** *Demansia rimicola* can be distinguished from other *Demansia* species within its range by the pale and dark lateral stripes. **Conservation.** IUCN status: Least Concern. No known major threats. Cats are recorded to prey on this species.

Blacksoil Whipsnake (*Demansia rimicola*), Hamilton Channels, QLD

Rufous Whipsnake
Demansia rufescens

TL 670 mm. **Lethality.** Potentially dangerous. **Description.** Body slender with a long slender tail. Reddish brown to coppery brown with a weakly reticulated pattern; olive gray to dark gray on head and neck. Head long, with moderately pointed snout, distinctive dark comma-shaped mark from eye to corner of mouth, prominent pale bar in front of and behind eye, and pale-edged dark line across snout. Ventral area white or cream. **Scalation.** Dorsal scales smooth and in 15 rows at mid-body. Ventrals 177–200. Anal scale divided. Subcaudals 65–85, all divided. **Habitat and rang.** Occurs on spinifex-dominated stony ground in arid scrublands and woodlands in the Pilbara region of WA. Also recorded on Barrow and Dolphin Islands, WA. **Behavior.** Diurnal. An active, fast-moving species generally observed foraging throughout the day. Clutch size not known. **Identification.** *Demansia rufescens* has a small distribution and can be distinguished from other *Demansia* species within its range by its overall reddish coloration and olive gray to dark gray head and neck. **Conservation.** IUCN status: Least Concern. No known major threats.

Rufous Whipsnake (*Demansia rufescens*), Indee, WA

Shine's Whipsnake
Demansia shinei

TL 810 mm. **Lethality.** Potentially dangerous. **Description.** Body slender with a long slender tail. Gray to brown with light brown head and yellowish flush on tail. Head long, with moderately pointed snout, distinctive dark comma-shaped mark from eye to corner of mouth, and prominent pale bar in front of and behind eye. Prominent brown band across nape and across rear of head is edged with yellow or orange. Ventral area pale yellow. **Scalation.** Dorsal scales smooth and in 15 rows at mid-body. Ventrals 177–207. Anal scale divided. Subcaudals 69–99, all divided. **Habitat and range.** Occurs in arid to semiarid spinifex-dominated landscapes, rocky hills, and sand ridges from central and s. NT into WA, with a possibly isolated population in e. Pilbara. **Behavior.** Diurnal. An active, fast-moving species generally observed foraging throughout the day. Clutch size not known. **Identification.** *Demansia shinei* can be distinguished from other *Demansia* species within its range by the pale head and the brown bands on head and nape edged in yellow or orange. **Conservation.** IUCN status: Least Concern. No known major threats.

Shine's Whipsnake (*Demansia shinei*), Tennant Creek area, NT

Shine's Whipsnake (*Demansia shinei*), juvenile, Tennant Creek area, NT

Gray Whipsnake
Demansia simplex

TL 530 mm. **Lethality.** Potentially dangerous. **Description.** Body slightly robust with a slender tail. Uniform brown to gray with the vertebral zone slightly darker than the flanks. Head long, with moderately pointed snout, distinctive dark comma-shaped mark from eye to corner of mouth, and creamy-yellow margin around eye. Ventral area white to cream. **Scalation.** Dorsal scales smooth and in 15 rows at mid-body. Ventrals 140–150. Anal scale divided. Subcaudals 55–65, all divided. **Habitat and range.** Occurs in open subhumid grassy woodlands of the Kimberley region of WA and the nw. NT. **Behavior.** Diurnal. An active, fast-moving species generally observed foraging throughout the day. Clutch size not known. **Identification.** *Demansia simplex* has a small distribution and can be identified by its uniform coloration and fewer than 160 ventral scales. **Conservation.** IUCN status: Least Concern. No known major threats.

Gray Whipsnake (*Demansia simplex*), Doongan Station, WA

Collared Whipsnake
Demansia torquata

TL 825 mm. **Lethality.** Potentially dangerous. **Description.** Body slender with a long slender tail. Brown to bluish gray, with top of head darker, ranging to black, and a broad, pale-edged dark bar across nape. Head long, with moderately pointed snout, distinctive dark comma-shaped mark from eye to corner of mouth, pale-edged bar in front of and behind eye, and pale-edged dark line across snout. Ventral area reddish. **Scalation.** Dorsal scales smooth and in 15 rows at mid-body. Ventrals 185–220. Anal scale divided. Subcaudals 70–90, all divided. **Habitat and range.** Occurs in a variety of habitats from tropical savannas and open woodlands to rain-forest edges in the coast and ranges of e. QLD from Gladstone to Cape York Peninsula. Also found on many islands in the Whitsunday group, QLD. **Behavior.** Diurnal. An active, fast-moving species generally observed foraging throughout the day. Clutches of 5–7 eggs recorded. **Identification.** *Demansia torquata* can be distinguished from other *Demansia* species within its range by the broad, pale-edged dark bar across its nape. **Conservation.** IUCN status: Least Concern. No known major threats.

Collared Whipsnake (*Demansia torquata*), Conway National Park, QLD

Genus *Denisonia*
De Vis' Banded and Ornamental Snakes

The two species of *Denisonia* are small to medium-size, robust snakes restricted to e. Australia and associated with woodlands and shrublands, mostly in areas subjected to seasonal flooding. They have weakly glossy, smooth scales in 17 mid-body rows and a relatively flat head with large eyes and a vertically elliptical pupil. Secretive, nocturnal snakes, they shelter in cracking clay soils. Diet consists primarily of frogs. They are live-bearing.

Both species become defensive if disturbed, flattening the body, thrashing wildly, and striking. A recorded bite from an Ornamental Snake (*Denisonia maculata*) produced severe swelling and discoloration of the hand.

DeVis' Banded Snake
Denisonia devisi

TL 570 mm. **Lethality.** Potentially dangerous. **Description.** Body medium-size and robust with a short tail. Light brown to yellowish brown or olive, with numerous irregular darker bands breaking up into blotches and head dark brown with paler brown flecks. Head broad, flat, and distinct from neck, with moderately rounded snout, and lips with conspicuous cream and dark brown bars. Ventral area white or cream. **Scalation.** Dorsal scales smooth and in 17 rows at mid-body. Ventrals 120–150. Anal scale single. Subcaudals 20–40, all single. **Habitat and range.** Occurs in grassy floodplains, woodlands, and shrublands through inland areas of QLD and NSW. Also recorded in riverine habitats in the sw. Murray-Darling Basin, in NSW, VIC, and SA.

Behavior. Nocturnal. A cryptozoic snake usually observed foraging on warm nights. Prefers moist areas, sheltering beneath ground debris and in soil cracks. Litters of about 5 young recorded. Bites readily if provoked. **Identification.** The distinctive pattern and separate distribution distinguish *Denisonia devisi* from *D. maculata*, which has a plain pattern. It has been confused with the Common Death Adder (*Acanthophis antarcticus*), which can be distinguished by its 21–23 mid-body scale rows, large triangular head, and tail with a terminal soft spine. **Conservation.** IUCN status: Least Concern. Not considered to be undergoing significant declines.

DeVis' Banded Snake (*Denisonia devisi*), Ned's Corner, VIC

DeVis' Banded Snake (*Denisonia devisi*)

Ornamental Snake
Denisonia maculata

TL 465 mm. **Lethality.** Potentially dangerous. **Description.** Body small and robust with a short tail. Brownish gray to dark gray with little pattern; head with darker patch. Head broad, flat, and distinct from neck, with moderately rounded snout, and lips with conspicuous black and white bars. Ventral area gray with dark flecking. **Scalation.** Dorsal scales smooth and in 17 rows at mid-body. Ventrals 120–150. Anal scale single. Subcaudals 20–40, all single. **Habitat and range.** Occurs in grassy floodplains and woodlands in the Dawson River drainage and Bowen Basin area of central e. QLD. **Behavior.** Nocturnal. A cryptozoic snake usually observed foraging on warm nights. Inhabits moist areas, sheltering in deep cracking clay soils. Litters of 3–11 young recorded. **Identification.** *Denisonia maculata* is distinguished from *D. devisi* by its plain pattern and separate distribution. It could be confused with the Common Death Adder (*Acanthophis antarcticus*), which can be distinguished by its 21–23 mid-body scale rows, large triangular head, and tail with a terminal soft spine. **Conservation.** IUCN status: Data Deficient; previously assessed as Vulnerable. No known major threats, but habitat conversion to agriculture is a threat in parts of the range.

Ornamental Snake (*Denisonia maculata*), The Willows, QLD

Genus *Drysdalia*
White-lipped, Masters', and Mustard-bellied Snakes

This genus contains three species of small, moderately robust snakes that occur in e. and s. Australia and are associated with dry woodlands, open forests, and heaths. They have matte-textured, smooth scales in 15 mid-body rows, and a moderately pointed snout. Diurnal snakes, they are often observed basking among tussocks at the edge of forest tracks. Diet consists primarily of skinks. They are live-bearing. Bites are recorded to cause significant swelling and pain.

White-lipped Snake
Drysdalia coronoides

TL 450 mm. **Lethality.** Potentially dangerous. **Description.** Body small and slightly robust with a short tail. Color variable, ranging from olive green to reddish brown or almost black. Head not distinct from neck, with moderately pointed snout, and upper lip with conspicuous white stripe. Ventral area cream to salmon pink. **Scalation.** Dorsal scales smooth, matte-textured, and in 15 rows at mid-body. Ventrals 120–160. Anal scale single. Subcaudals 35–70, all single. **Habitat and range.** Associated mostly with dry woodlands and tussock grasses in open forests throughout se. Australia and TAS, including higher alpine areas. **Behavior.** Diurnal. Usually observed basking among grass tussocks along tracks. Litters of 2–10 young recorded. **Identification.** *Drysdalia coronoides* is similar to *D. mastersii* and *D. rhodogaster* but lacks a band across the nape. It could be confused with the Marsh Snake (*Hemiaspis signata*) in ne. NSW, but that species has 17 mid-body scale rows and a pale stripe behind the eye in addition to one on the upper lip. **Conservation.** IUCN status: Least Concern. No known major threats. Habitat loss and predation by cats are localized threats.

White-lipped Snake (*Drysdalia coronoides*), Ballarat, VIC

Masters' Snake
Drysdalia mastersii

TL 330 mm. **Lethality.** Potentially dangerous. **Description.** Body small and slightly robust with a short tail. Yellowish brown to gray with yellow to whitish band across nape. Head darker than body, not distinct from neck, with moderately pointed snout and conspicuous white stripe on upper lip. Ventral area orange with dark spots and flecks. **Scalation.** Dorsal scales smooth, matte-textured, and in 15 rows at mid-body. Ventrals 140–160. Anal scale single. Subcaudals 40–55, all single. **Habitat and range.** Inhabits coastal dunes, limestones, and mallee woodlands with heath and spinifex in nw. VIC and parts of s. SA and WA. **Behavior.** Diurnal. Usually observed basking among vegetation or found sheltering beneath ground debris. Litters of 2 or 3 young recorded. **Identification.** *Drysdalia mastersii* is similar to *D. coronoides* and *D. rhodogaster* but has a separate distribution. **Conservation.** IUCN status: Least Concern. No known major threats. Habitat loss and predation by cats are localized threats.

Masters' Snake (*Drysdalia mastersii*), Pinnaroo district, SA

Mustard-bellied Snake
Drysdalia rhodogaster

TL 400 mm. **Lethality.** Potentially dangerous. **Description.** Body small and slightly robust with a short tail. Brownish to gray with fine, darker flecks and a conspicuous yellow to orange band across nape. Head dark brown to gray, not distinct from neck, with moderately pointed snout, and lips mottled with gray or black. Ventral area yellow to pink. **Scalation.** Dorsal scales smooth, matte-textured, and in 15 rows at mid-body. Ventrals 140–160. Anal scale single. Subcaudals 40–55, all single. **Habitat and range.** Mostly associated with dry woodlands, heaths, and tussock grasses in open forests east of the Great Dividing Range from the central coast region of NSW south to the VIC border. **Behavior.** Diurnal. Usually observed basking among grass tussocks or sheltering beneath ground debris. Litters of 2–6 young recorded. **Identification.** *Drysdalia rhodogaster* is similar to *D. mastersii* but has a completely separate distribution. It is distinguished from *D. coronoides* by the conspicuous band across the nape. **Conservation.** IUCN status: Least Concern. No known major threats. Habitat loss and predation by cats are localized threats.

Mustard-bellied Snake (*Drysdalia rhodogaster*), Blue Mountains, NSW

Genus *Echiopsis*
Bardick

The single species of *Echiopsis* is a medium-size, stout snake occurring across s. Australia that is associated mostly with semiarid woodlands, mallee, and heath. It has smooth, matte-textured scales in 17–21 mid-body rows, a broad head, distinct from the neck, and moderately large eyes. Diurnal and nocturnal, it is a sedentary snake sometimes observed basking. Diet consists primarily of skinks. It is livebearing. This snake is pugnacious when disturbed, and a bite may cause severe symptoms.

Bardick
Echiopsis curta

TL 710 mm. **Lethality.** Potentially dangerous. **Description.** Body medium-size and robust with a short tail. Dark gray, olive brown, or reddish brown with paler lateral scales. Head slightly darker than body, distinct from neck, with moderately rounded snout, and lips flecked with white. Ventral area pale gray-brown to yellowish under tail. **Scalation.** Dorsal scales smooth, matte-textured, and in 17–21 (usually 19) rows at mid-body. Ventrals 120–145. Anal scale single. Subcaudals 25–45, all single. **Habitat and range.** Associated with semiarid mallee woodlands in nw. VIC and adjacent SA, the Eyre Peninsula, and sw. WA. **Behavior.** Diurnal but also nocturnal. May be observed by day basking among spinifex and other low vegetation. Litters of 8–10 young recorded. Bites readily if provoked. **Identification.** *Echiopsis curta* may be confused with the Common Death Adder (*Acanthophis antarcticus*), but that species has a soft terminal spine on the tail. **Conservation.** IUCN status: Least Concern. Threats include habitat loss to agriculture and inappropriate fire regimes; also, possible predation by foxes.

Bardick (*Echiopsis curta*), Pinnaroo district, SA

Genus *Elapognathus*
Crowned and Hedge Snakes

The two species of *Elapognathus* are small to medium-size, moderately robust or slender snakes that are restricted to sw. Australia. Associated mostly with coastal heaths, swamps, and woodlands, they shelter beneath ground debris. They have smooth, matte-textured scales in 15 mid-body rows and a moderately pointed snout. Diurnal snakes, they are usually observed basking. Diet consists primarily of frogs and skinks. They are live-bearers. *Elapognathus* species are generally inoffensive but may bite if handled.

Western Crowned Snake
Elapognathus coronatus

TL 600 mm. **Lethality.** Potentially dangerous. **Description.** Body medium-size and slender with a short tail. Olive gray to olive brown or tan, with silver to gray head and broad black band across nape. Head not distinct from neck, with moderately pointed snout and conspicuous, black-edged white streak on upper lip. Ventral area reddish orange to yellow. **Scalation.** Dorsal scales smooth, matte-textured, and in 15 rows at mid-body. Ventrals 130–160. Anal scale single. Subcaudals 35–50, all single. **Habitat and range.** Coastal woodlands, heaths, and swamps of s. and sw. WA. **Behavior.** Diurnal. Usually observed basking among vegetation. Shelters beneath ground debris and inhabits abandoned nests of stick-nest ants (*Iridomyrmex conifer*). Litters of 3–9 young recorded. **Identification.** *Elapognathus coronatus* can be distinguished from the similar *E. minor* by the broad black band across the nape. **Conservation.** IUCN status: Least Concern. Historical land clearing, including of wetlands, has occurred in parts of the range.

Western Crowned Snake (*Elapognathus coronatus*), D'Ebtrecasteaux National Park, WA

Short-nosed Snake
Elapognathus minor

TL 500 mm. **Lethality.** Potentially dangerous. **Description.** Body small and slightly robust with a short tail. Gray to reddish brown to bright red on tail. Skin between scales dark, producing reticulated pattern. Head not distinct from neck, with short snout; side of head white to pale gray, with prominent yellow to orange, oblique bar on neck. Ventral area dark yellow to orange laterally. **Scalation.** Dorsal scales smooth, matte-textured, and in 15 rows at mid-body. Ventrals 115–130. Anal scale single. Subcaudals 40–55, all single. **Habitat and range.** Occurs on sandy soil in heaths along margins of swamps with sedges, and in wet sclerophyll forests, from Two People's Bay northwest to Busselton, in sw. WA. **Behavior.** Diurnal. Shelters in low, dense vegetation. Litters of 8–12 young recorded. **Identification.** *Elapognathus minor* can be distinguished from the similar *E. coronatus* by the oblique orange-yellow bar on side of neck. **Conservation.** IUCN status: Least Concern. Despite historical habitat loss to land clearing, this species is unlikely to have any major threats.

Short-nosed Snake (*Elapognathus minor*), Rocky Gulley, WA

Genus *Furina*
Naped Snakes

The three species of small to medium-size, slender to moderately robust naped snakes occur in a variety of habitats throughout most of Australia and are associated mostly with woodlands, savannas, and rocky outcrops. They have glossy, smooth scales in 15–17 mid-body rows and a pale or bright band on the nape. Nocturnal, terrestrial snakes, they shelter beneath ground debris and in soil cracks. Diet consists primarily of skinks. Naped snakes are oviparous. Bites from larger specimens may cause severe symptoms.

Yellow-naped Snake
Furina barnardi

TL 500 mm. **Lethality.** Potentially dangerous. **Description.** Body small and moderately robust with a short tail. Dark gray-brown to black with pale-edged scales forming reticulated appearance. Head slightly distinct from neck, with squarish snout and broad pale brown to yellow collar on nape. Ventral area white or cream. **Scalation.** Dorsal scales smooth, glossy, and in 15 rows at mid-body. Ventrals 170–200. Anal scale divided. Subcaudals 25–45, all divided. **Habitat and range.** Inhabits rock outcrops and woodlands on the coast and interior of e. QLD from Port Curtis to the base of Cape York Peninsula. **Behavior.** Nocturnal. Shelters beneath fallen timber and other ground debris. Bites readily if disturbed. **Identification.** *Furina barnardi* can be distinguished from the similar *F. diadema* and *F. ornata* by the pale brown to yellow band on its nape. **Conservation.** IUCN status: Least Concern. No known substantial threats, but some habitat has been lost to intensive agricultural development.

Yellow-naped Snake (*Furina barnardi*), Marlborough area, QLD

Red-naped Snake
Furina diadema

TL 400 mm. **Lethality.** Potentially dangerous. **Description.** Body small and slender with a short tail. Reddish brown with dark-edged scales forming reticulated appearance. Head slightly distinct from neck, with squarish snout and bright orange-red blotch on nape. Ventral area white or cream. **Scalation.** Dorsal scales smooth, glossy, and in 15 rows at mid-body. Ventrals 160–210. Anal scale divided. Subcaudals 35–70, all divided. **Habitat and range.** Occurs in a variety of habitats but associated mostly with dry woodlands and grasslands throughout much of e. Australia. **Behavior.** Nocturnal. Shelters beneath ground debris during the day. Clutches of 1–5 eggs recorded. **Identification.** *Furina diadema* can be distinguished from the similar *F. barnardi* and *F. ornata* by the bright orange-red blotch on the nape and distribution. **Conservation.** IUCN status: Least Concern. Vulnerable to habitat modification caused by urbanization in parts of its range.

Red-naped Snake (*Furina diadema*), Berajondo, QLD

Orange-naped Snake
Furina ornata

TL 650 mm. **Lethality.** Potentially dangerous. **Description.** Body medium-size and slender with a short tail. Reddish brown to dark brown with dark-edged scales forming reticulated appearance. Head slightly distinct from neck, with squarish snout and bright orange-red band on nape. Ventral area white. **Scalation.** Dorsal scales smooth, glossy, and in 15–17 rows at mid-body. Ventrals 160–240. Anal scale divided. Subcaudals 35–70, all divided. **Habitat and range.** Occurs in a variety of habitats from dry tropical woodlands to desert grasslands throughout much of n. and w. Australia. **Behavior.** Nocturnal. Shelters beneath ground debris during the day. Clutches of 3–6 eggs recorded. **Identification.** *Furina ornata* can be distinguished from the similar *F. barnardi* and *F. diadema* by the bright red to orange band on the neck and its northern and western distribution. **Conservation.** IUCN status: Least Concern. No known threats.

Orange-naped Snake (*Furina ornata*), Warburton, WA

Genus *Hemiaspis*
Gray Snake and Swamp Snake

This genus contains two species of medium-size, robust snakes found in e. Australia. One species occurs in cracking soils on floodplains, and the other is associated with moist environments of the coast and ranges. They have weakly glossed, smooth scales in 17 mid-body rows and a slightly rounded snout. One species is diurnal, feeding on frogs and skinks, the other nocturnal, feeding only on frogs. Both are live-bearers. Bites from larger specimens may cause severe symptoms.

Gray Snake
Hemiaspis damelii

TL 600 mm. **Lethality.** Potentially dangerous. **Description.** Body medium-size and slender with a slender tail. Olive gray to dark gray with lateral scales tipped anteriorly with black. Head not distinct from neck, with slightly rounded snout, and black band on base of head and nape. Ventral area yellowish cream or white. **Scalation.** Dorsal scales smooth and in 17 rows at mid-body. Ventrals 140–170. Anal scale divided. Subcaudals 35–50, all single. **Habitat and range.** Occurs in cracking soils on floodplains within dry sclerophyll forests and woodlands from central NSW north to QLD around Rockhampton. Also, old records exist from the junction of the Lachlan and Murrumbidgee Rivers in NSW. **Behavior.** Crepuscular and nocturnal. Shelters beneath ground debris and in soil cracks. Feeds exclusively on frogs. Litters of 6–12 young recorded. **Identification.** *Hemiaspis damelii* has a black band on the base of the head and usually a yellowish-cream or white ventral area. The similar *H. signata* has two narrow pale stripes on each side of the face and usually a dark gray to black ventral surface. **Conservation.** IUCN status: Endangered. Threats include land clearing for intensive agricultural use and hydrological changes to waterways within its habitat.

Gray Snake (*Hemiaspis damelii*), Narrabri, NSW

Marsh Snake
Hemiaspis signata

TL 700 mm. **Lethality.** Potentially dangerous. **Description.** Body medium-size and slender with a slender tail. Brown to dark olive gray, with head sometimes darker. Head not distinct from neck, with slightly rounded snout, two narrow pale stripes on each side of face, and a pale stripe behind the eye and another on the upper lip. Ventral area usually gray to black. **Scalation.** Dorsal scales smooth and in 17 rows at mid-body. Ventrals 150–170. Anal scale divided. Subcaudals 40–60, all single. **Habitat and range.** Usually encountered around watercourses in wetter forests along the coast and ranges of e. Australia from se. NSW to parts of ne. QLD, where there are some isolated populations. **Behavior.** Diurnal, crepuscular, and nocturnal. Feeds on skinks and frogs. Litters of 4–20 young recorded. **Identification.** *Hemiaspis signata* has two narrow pale stripes on each side of face and usually a dark gray to black ventral surface. The similar *H. damelii* has a black band on the base of the head and usually a yellowish-cream or white ventral surface. **Conservation.** IUCN status: Least Concern. No known threats.

Marsh Snake (*Hemiaspis signata*), Mt. Nebo, QLD

Genus *Neelaps*
Burrowing Snakes

The two species of small snakes in this genus are restricted to s. SA and WA. Associated with dry, sandy woodlands and heaths, they shelter in leaf litter. Both have glossy, smooth scales in 15 mid-body rows, a flat head, and a rounded snout. Secretive burrowing snakes, they are usually encountered at night. Diet consists of small skinks. They are oviparous. These burrowers have a small mouth and are disinclined to bite.

Black-naped Burrowing Snake
Neelaps bimaculatus

TL 450 mm. **Lethality.** Potentially dangerous; unlikely to cause significant envenomation. **Description.** Body small and slender with a short tail. Orange to reddish brown with darker-edged mid-body scales forming a reticulated pattern, black mark on snout, and a broad black band on head and another on nape. Head narrow and flat, with rounded, protrusive snout. Ventral area cream to white. **Scalation.** Dorsal scales smooth, glossy, and in 15 rows at mid-body. Ventrals 176–228. Anal scale divided. Subcaudals 19–30, divided. **Habitat and range.** Occurs in mallee woodlands, coastal heaths, and sand dunes in sw. Australia, and an isolated population in spinifex at North West Cape, WA. Also found on the Eyre Peninsula and near Nullarbor in SA. **Behavior.** Nocturnal. A burrowing snake that dwells in leaf litter and is usually encountered on the surface only at night. Clutches of 2–6 eggs recorded. **Identification.** *Neelaps bimaculatus* can be distinguished from *N. calonotos* by its lack of a black vertebral stripe. **Conservation.** IUCN status: Least Concern. No known major threats.

Black-naped Burrowing Snake (*Neelaps bimaculatus*), Perth, WA

Black-striped Burrowing Snake
Neelaps calonotos

TL 280 mm. **Lethality.** Potentially dangerous; unlikely to cause significant envenomation. **Description.** Body small and slender with a short tail. Orange to reddish brown with individual cream scales, a black vertebral stripe enclosing white spots from nape to tail tip, black mark on snout, and a broad black band on head and another on nape. Head narrow and flat, with rounded, protrusive snout. Ventral area cream to white. **Scalation.** Dorsal scales smooth, glossy, and in 15 rows at mid-body. Ventrals 126–143. Anal scale divided. Subcaudals 23–35, divided. **Habitat and range.** Occurs in coastal heaths, woodlands, and sand dunes in a small area between Mandurah and Dongara in sw. WA. **Behavior.** Nocturnal. A burrowing snake that dwells in leaf litter and is usually encountered on the surface only at night. Clutches of 2–5 eggs recorded. **Identification.** *Neelaps calonotos* is distinguished from *N. bimaculatus* by the white-spotted black vertebral stripe from nape to tail tip. **Conservation.** IUCN status: Near Threatened. Seriously threatened by increasing urban development and possibly from fire within its restricted range.

Black-striped Burrowing Snake (*Neelaps calonotos*), Lancelin, WA

Genus *Rhinoplocephalus*
Square-nosed Snake

This genus contains one species, a small, relatively robust snake restricted to sw. Australia. It shelters in disused nests of stick-nest ants and beneath grass trees (*Xanthorrhoea*) in dry woodlands and scrublands. It has glossy, smooth scales in 15 mid-body rows and a squarish snout. Predominantly nocturnal, it is also observed basking. Diet consists of skinks. It is live-bearing. This snake is not inclined to bite.

Square-nosed Snake
Rhinoplocephalus bicolor

TL 450 mm. **Lethality.** Potentially dangerous. **Description.** Body moderately robust with a short tail. Dark brown to dark gray with orange on the mid- to lower flanks. Head depressed, with squarish snout. Ventral area creamy white. **Scalation.** Dorsal scales smooth, glossy, and in 15 rows at mid-body. Ventrals 135–165. Anal scale single. Subcaudals 20–45, all single. **Habitat and range.** Coastal dunes, heaths, and woodlands in the far sw. corner of WA, from Cape Arid National Park to Busselton. **Behavior.** Mostly nocturnal, though sometimes observed basking in vegetation. Shelters beneath rocks, logs, and other ground debris. Litters of 1–5 young recorded. **Identification.** *Rhinoplocephalus bicolor* has a very restricted distribution and can be identified by its unusual coloration, depressed head, and squarish snout. **Conservation.** IUCN status: Least Concern. No known major threats.

Square-nosed Snake (*Rhinoplocephalus bicolor*), Yelverton, WA

Genus *Simoselaps*
Burrowing Snakes

The four species of *Simoselaps* are small, moderately robust snakes that occur in SA, WA, and the NT. They are associated with sandy areas in dry woodlands, heaths, coastal dunes, and spinifex deserts. Most species are distinctly banded. All have glossy, smooth scales in 15 mid-body rows and a protruding, flattened snout. Secretive burrowing snakes, they are usually encountered at night. Diet consists of skinks. They are oviparous. These burrowers have a small mouth and are not inclined to bite.

Desert Banded Snake
Simoselaps anomalus

TL 210 mm. **Lethality.** Potentially dangerous; unlikely to cause significant envenomation. **Description.** Body small and moderately robust with a short tail. Reddish orange to yellowish with 24–40 black bands, equal to or narrower than the pale interspaces, between nape and tail tip. Head depressed and glossy black, with white band behind head and protruding, flattened white snout. Ventral area creamy white. **Scalation.** Dorsal scales smooth, glossy, and in 15 rows at mid-body. Ventrals 115–130. Anal scale divided. Subcaudals 15–30, divided. **Habitat and range.** Occurs in sandy soils with spinifex and mulga woodlands from nw. coastal areas of WA eastward to sw. NT and nw. SA. **Behavior.** Nocturnal. A burrowing snake usually encountered on the surface only at night. Shelters beneath ground cover and in leaf litter. Clutches of 2 or 3 eggs recorded. **Identification.** *Simoselaps anomalus* can be distinguished from other *Simoselaps* species by its glossy black head and 24–40 black bands. **Conservation.** IUCN status: Least Concern. No known major threats.

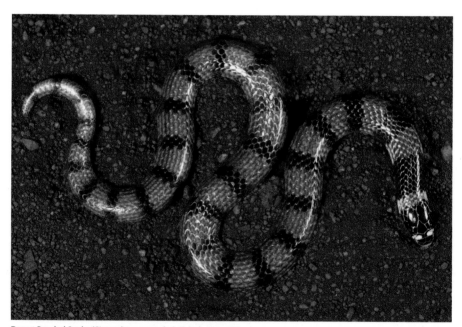

Desert Banded Snake (*Simoselaps anomalus*), Kaltukatjara, NT

Jan's Banded Snake
Simoselaps bertholdi

TL 300 mm. **Lethality.** Potentially dangerous; unlikely to cause significant envenomation. **Description.** Body small and moderately robust with a short tail. Yellow to reddish orange with 18–31 black bands, approximately equal to the pale interspaces, between nape and tail tip. Scales with dark edges laterally. Head depressed and white to pale gray, with protruding flattened snout. Ventral area creamy white. **Scalation.** Dorsal scales smooth, glossy, and in 15 rows at mid-body. Ventrals 115–135. Anal scale divided. Subcaudals 15–30, divided. **Habitat and range.** Occurs in coastal dunes and heaths to arid shrublands and woodlands from central WA to w. SA and sw. NT. **Behavior.** Nocturnal. A burrowing snake usually encountered on the surface only at night. Shelters beneath ground cover and in leaf litter. Clutches of 1–8 eggs recorded. **Identification.** *Simoselaps bertholdi* can be distinguished from other *Simoselaps* species by its glossy white to pale gray head and 18–31 broad black bands. **Conservation.** IUCN status: Least Concern. No known major threats.

Jan's Banded Snake (*Simoselaps bertholdi*), Port Augusta area, SA

West Coast Banded Snake
Simoselaps littoralis

TL 390 mm. **Lethality.** Potentially dangerous; unlikely to cause significant envenomation. **Description.** Body small and moderately robust with a short tail. Yellow to yellowish white with 20–42 narrow black bands, usually narrower than the pale interspaces, between nape and tail tip. Scales without dark edges. Head depressed and white to pale gray, with black band behind head and protruding flattened snout. Ventral area creamy white. **Scalation.** Dorsal scales smooth, glossy, and in 15 rows at mid-body. Ventrals 100–125. Anal scale divided. Subcaudals 15–25, divided. **Habitat and range.** Occurs in coastal dunes, heathlands, and sand-ridge deserts from Cervantes to North West Cape, WA. **Behavior.** Nocturnal. A burrowing snake usually encountered on the surface only at night. Shelters beneath ground cover and in leaf litter. Clutches of 3 or 4 eggs recorded. **Identification.** *Simoselaps littoralis* can be distinguished from other *Simoselaps* species by its white to pale gray head and 20–42 narrow black bands. **Conservation.** IUCN status: Least Concern. No known major threats.

West Coast Banded Snake (*Simoselaps littoralis*), Shark Bay, WA

Dampierland Burrowing Snake
Simoselaps minimus

TL 215 mm. **Lethality.** Potentially dangerous; unlikely to cause significant envenomation. **Description.** Body small and moderately robust with a short tail. Yellowish brown with dark-edged scales forming a reticulated pattern between nape and tail tip. Head depressed and black, with protruding, flattened white snout, and black band on nape. Ventral area creamy white. **Scalation.** Dorsal scales smooth, glossy, and in 15 rows at mid-body. Ventrals 100–125. Anal scale divided. Subcaudals 19–25, mostly divided, 1 and 3–5 single. **Habitat and range.** Occurs in coastal dunes on the Dampier Peninsula, sw. Kimberley, WA. **Behavior.** Nocturnal. A poorly known burrowing snake usually encountered on the surface only at night. Shelters beneath ground cover and in leaf litter. Presumably oviparous. **Identification.** *Simoselaps minimus* has a very restricted distribution and can be distinguished from other *Simoselaps* species by its reticulated pattern and lack of bands along body. **Conservation.** IUCN status: Least Concern. No known major threats.

Dampierland Burrowing Snake (*Simoselaps minimus*), Broome, WA

Genus *Suta*
Hooded Snakes

See complete genus description in the Medically Significant Venomous Land Snakes section (p. 84). The nine species listed here are potentially dangerous. A bite from a Little Whip Snake (*S. flagellum*) in 2007 caused the death of an adult; an anaphylactic reaction may have been involved. All species should be treated as potentially dangerous. Bites are recorded to cause headaches, swelling, and nausea.

Dwyer's Snake
Suta dwyeri

TL 600 mm. **Lethality.** Potentially dangerous. **Description.** Body medium-size and moderately robust with a short tail. Yellow-brown to reddish brown with dark-edged scales that form dark vertebral zone. Head slightly depressed, not distinct from neck, with slightly rounded snout, black head blotch unbroken from snout to nape, and noticeable pale preocular blotch. Ventral area white. **Scalation.** Dorsal scales smooth, glossy, and in 15 rows at mid-body. Ventrals 135–170. Anal scale single. Subcaudals 20–40, all single. **Habitat and range.** Associated with rock outcrops in dry woodlands and mallee areas of QLD, NSW, and VIC, and as far west as the Eyre Peninsula in SA. **Behavior.** Nocturnal. A secretive species that shelters beneath rock slabs and other ground debris. Average litters of 3 young recorded. **Identification.** *Suta dwyeri* can be identified by the distinctive black head blotch, unbroken from snout to nape, and noticeable pale preocular blotch. **Conservation.** IUCN status: Least Concern. No known threats.

Dwyer's Snake (*Suta dwyeri*), Bendigo, VIC

Rosen's Snake
Suta fasciata

TL 620 mm. **Lethality.** Potentially dangerous. **Description.** Body medium-size and moderately robust with a short tail. Reddish brown to orange, yellowish, or gray, with many dark bands and blotches. Head slightly depressed, not distinct from neck, with slightly rounded snout, and dark streak extending from nostril through eye to side of neck. Ventral area creamy white. **Scalation.** Dorsal scales smooth, glossy, and in 17–19 rows at midbody. Ventrals 140–165. Anal scale single. Subcaudals 20–40, all single. **Habitat and range.** Associated with sandy and stony soils in woodlands and arid shrublands in central WA. **Behavior.** Nocturnal. A secretive species that shelters beneath leaf litter and ground debris. Litters of 1–7 young recorded. **Identification.** *Suta fasciata* can be identified by its prominently banded and blotched coloration and a distribution restricted to central WA. **Conservation.** IUCN status: Least Concern. No known major threats.

Rosen's Snake (*Suta fasciata*), Laverton, WA

Little Whip Snake
Suta flagellum

TL 450 mm. **Lethality.** Potentially dangerous. **Description.** Body small and moderately robust with a short tail. Light tan to brown or orange, with dark-edged scales that form reticulated pattern. Head slightly depressed, not distinct from neck, with slightly rounded snout and black head blotch broken by pale bar across snout. Ventral area cream to brown. **Scalation.** Dorsal scales smooth, glossy, and in 17 rows at mid-body. Ventrals 125–150. Anal scale single. Subcaudals 20–40, all single. **Habitat and range.** Occurs in granite outcrops in dry woodlands and basalt grassland plains in se. SA, s. VIC, and se. NSW. **Behavior.** A secretive nocturnal species that shelters beneath rock slabs and other ground debris. Litters of 2–11 young recorded.

Identification. *Suta flagellum* can be identified by the distinct black head blotch broken by a pale bar across the snout and 17 mid-body scale rows. **Conservation.** IUCN status: Least Concern. Threatened in part of its range from predation by feral animals and habitat modification for agriculture.

Little Whip Snake (*Suta flagellum*), Skipton, VIC

Pilbara Hooded Snake
Suta gaikhorstorum

TL 460 mm. **Lethality.** Potentially dangerous. **Description.** Body small and moderately robust with a short tail. Pale reddish brown to bright red, with little or no dark edges to scales. Head slightly depressed, not noticeably distinct from neck, with slightly rounded snout and a black head blotch, unbroken from snout to nape. Ventral area white. **Scalation.** Dorsal scales smooth, glossy, and in 15 rows at mid-body. Ventrals 160–168. Anal scale single. Subcaudals 23–34, all single. **Habitat and range.** Associated with stony soils and spinifex in the Pilbara region of WA. **Behavior.** A recently described species, it is presumably similar to other members of the genus. **Identification.** *Suta gaikhorstorum* can be identified by the distinct black head blotch, unbroken from snout to nape, no pale markings in front of eyes, and a restricted distribution. **Conservation.** IUCN status: Data Deficient. No known major threats.

Pilbara Hooded Snake (*Suta gaikhorstorum*), Hamersley Ranges, WA

Gould's Hooded Snake
Suta gouldii

TL 525 mm. **Lethality.** Potentially dangerous. **Description.** Body medium-size and moderately robust with a short tail. Orange-brown to grayish, with dark-edged scales that form a slight reticulated pattern. Head slightly depressed, not distinct from neck, with slightly rounded snout, black head blotch unbroken from snout to nape, and noticeable pale preocular mark. Ventral area white to cream. **Scalation.** Dorsal scales smooth, glossy, and in 15 rows at mid-body. Ventrals 150–180. Anal scale single. Subcaudals 25–40, all single. **Habitat and range.** Associated with heaths, woodlands, and shrublands of sw. WA. **Behavior.** Nocturnal. A secretive species. Shelters beneath rock slabs and other ground debris. Average litters of 1–6 young recorded. **Identification.** *Suta gouldii* can be identified by the distinct black head blotch, unbroken from snout to nape, the noticeable pale preocular mark, and its distribution, restricted to sw. Australia. **Conservation.** IUCN status: Least Concern. No known major threats.

Gould's Hooded Snake (*Suta gouldii*), Perth Hills, WA

Monk Snake
Suta monachus

TL 460 mm. **Lethality.** Potentially dangerous. **Description.** Body small and moderately robust with a short tail. Orange-brown to orange-red, sometimes with dark-edged scales that form a reticulated pattern. Head slightly depressed, not distinct from neck, with slightly rounded snout and a black head blotch, unbroken from snout to nape. Ventral area white. **Scalation.** Dorsal scales smooth, glossy, and in 15 rows at mid-body. Ventrals 150–180. Anal scale single. Subcaudals 21–35, all single. **Habitat and range.** Associated with rocky outcrops or hard red soils in acacia woodlands and shrublands from sw. NT and central SA through arid regions to coastal areas of WA. **Behavior.** A secretive nocturnal species that shelters beneath rock slabs and other ground debris. Average litters of 1–5 young recorded. **Identification.** *Suta monachus* can be identified by the distinct black head blotch, unbroken from snout to nape, lack of pale preocular markings, and its distribution. **Conservation.** IUCN status: Least Concern. No known threats.

Monk Snake (*Suta monachus*), Laverton, WA

Mitchell's Short-tailed Snake
Suta nigriceps

TL 590 mm. **Lethality.** Potentially dangerous. **Description.** Body medium-size and robust with a short tail. Reddish brown to purplish brown with conspicuous dark vertebral stripe or zone. Head slightly depressed, not distinct from neck, with slightly rounded snout; a black head blotch, unbroken from snout to nape and continuous with prominent dark vertebral coloring; and lips pale orange to white. Ventral area white. **Scalation.** Dorsal scales smooth, glossy, and in 15 rows at mid-body. Ventrals 145–175. Anal scale single. Subcaudals 18–35, all single. **Habitat and range.** Associated with mallee woodlands, granite outcrops, and sand plains of sw. WA. **Behavior.** Nocturnal. A cryptic species that shelters beneath logs and other ground debris. Average litters of 4 young recorded. **Identification.** *Suta nigriceps* can be identified by the distinct black head blotch, unbroken from snout to nape and continuous with prominent dark vertebral stripe or zone. It is restricted to sw. Australia. **Conservation.** IUCN status: Least Concern. Habitat loss and degradation caused by urbanization may present localized threats.

Mitchell's Short-tailed Snake (*Suta nigriceps*), Yanchep, WA

Ord Curl Snake
Suta ordensis

TL 760 mm. **Lethality.** Potentially dangerous. **Description.** Body medium-size and moderately robust with a short tail. Yellow-brown to dark brown or gray. Head slightly depressed, not noticeably distinct from neck, with slightly rounded snout; obscure dark hood, unbroken from snout to nape; and sometimes light barring on lips. Ventral area white to cream. **Scalation.** Dorsal scales smooth, glossy, and in 19 rows at mid-body. Ventrals 165–185. Anal scale single. Subcaudals 30–40, all single. **Habitat and range.** Occurs in tropical woodlands and grasslands in the Ord and Victoria river catchments of nw. WA. **Behavior.** Nocturnal. A secretive species, it is considered to have similar habits to the Curl Snake (*Suta suta*; p. 106). **Identification.** *Suta ordensis* can be distinguished from other *Suta* species by the obscure dark hood and lack of lateral light streak from nostril to temporal region. It also has a very restricted distribution. **Conservation.** IUCN status: Least Concern. No clear threats, but possibly at risk from consumption of Cane Toads and alteration of its habitat.

Ord Curl Snake (*Suta ordensis*), Kalkarinji, NT

Mallee Black-headed Snake
Suta spectabilis

TL 400 mm. **Lethality.** Potentially dangerous. **Description.** Body small and moderately robust with a short tail. Gray-brown to reddish brown with dark-edged scales that form a reticulated pattern. Head slightly depressed, not distinct from neck, with slightly rounded snout and a black head blotch, broken by pale bar across snout. Ventral area white. **Scalation.** Dorsal scales smooth, glossy, and in 15 rows at mid-body. Ventrals 135–170. Anal scale single. Subcaudals 20–40, all single. **Habitat and range.** Occurs in mallee woodlands and heathlands of the Big Desert in VIC; across s. SA, with an outlying population near Coober Pedy; and into se. WA. Also recorded around Broken Hill, NSW. **Behavior.** Nocturnal. A cryptic species that shelters beneath ground debris. Average litters of 3 young recorded. **Identification.** *Suta spectabilis* can be identified by the black head blotch broken by a pale bar across the snout and 15 mid-body scale rows. **Conservation.** IUCN status: Least Concern. No known threats.

Mallee Black-headed Snake (*Suta spectabilis*), Coober Pedy, SA

Genus *Vermicella*
Bandy Bandys

Six species of small to medium-size, long-bodied snakes, bandy bandys are distributed throughout Australia in a variety of habitats from the moist coast and ranges to the deserts of the interior. All have glossy, smooth scales in 15 mid-body rows, distinct black and white banding, and a rounded snout. Secretive burrowing snakes, they are usually encountered at night. Diet consists of blind snakes. They are oviparous.

Vermicella species have a small mouth and are disinclined to bite, though bites can cause pain, swelling, and systemic effects.

Common Bandy Bandy
Vermicella annulata

TL 760 mm. **Lethality.** Potentially dangerous. **Description.** Body medium-size and robust with a short, blunt-tipped tail. Dorsum with broad, alternating black and white rings along entire body and tail; 36–38 white rings. Head short, not distinct from neck, with rounded snout. Ventral area with black and white rings. **Scalation.** Dorsal scales smooth, glossy, and in 15 rows at mid-body. Ventrals 180–260. Anal scale divided. Subcaudals 10–35, all divided. **Habitat and range.** Occurs in a variety of habitats from wet and dry sclerophyll forests to mallee woodlands and arid sandy regions from Port Augusta, SA, through n. VIC, most of NSW, and QLD to ne. NT. **Behavior.** Nocturnal. A widespread burrowing species observed active on the surface only at night, often after rainstorms. Shelters beneath logs and rocks. Clutches of 4–6 eggs recorded. Raises parts of its body in loops off the ground if disturbed. **Identification.** *Vermicella annulata* is readily identified by its distinctive body shape and coloration, with 36–38 white rings. It mostly does not occur with other bandy bandy species, overlapping only with *V. parscauda* on the tip of Cape York Peninsula, QLD, but that species has 51–89 much narrower white rings. **Conservation.** IUCN status: Least Concern. No known major threats.

Common Bandy Bandy (*Vermicella annulata*), Oakey, QLD. Insert: Common Bandy Bandy (*Vermicella annulata*), head, Murray-Sunset National Park, VIC.

Intermediate Bandy Bandy
Vermicella intermedia

TL 605 mm. **Lethality.** Potentially dangerous. **Description.** Body medium-size and robust with a short, blunt-tipped tail. Dorsum with broad, alternating black and white rings along entire body and tail; 50–53 white rings. Head short, not distinct from neck, with rounded snout. Ventral area with black and white rings. **Scalation.** Dorsal scales smooth, glossy, and in 15 rows at mid-body. Ventrals 246–256. Anal scale divided. Subcaudals 15–28, all divided. **Habitat and range.** Occurs in tropical savannas and rocky vine thickets in the Top End of the NT across to the Kimberley, n. WA. **Behavior.** Nocturnal. A burrowing species observed active on the surface only at night, often after rainstorms. Shelters beneath logs and rocks. Raises parts of its body in loops off the ground if disturbed. **Identification.** *Vermicella intermedia* is readily identified by its distinctive body shape and coloration, with 50–53 white rings. It overlaps in distribution with *V. multifasciata*, which has 77–109 white rings that in some specimens are formed by white spots arranged in a band. **Conservation.** IUCN status: Least Concern. No known major threats.

Intermediate Bandy Bandy (*Vermicella intermedia*), Adelaide River, NT

Northern Bandy Bandy
Vermicella multifasciata

TL 360 mm. **Lethality.** Potentially dangerous. **Description** .Body small and robust with a short, blunt-tipped tail. Black to dark brown with dark and white rings along entire body and tail; 77–109 white rings, which in some specimens are formed by numerous white spots arranged in bands. Head short, not distinct from neck, with rounded snout. Ventral area with dark and white rings. **Scalation.** Dorsal scales smooth, glossy, and in 15 rows at mid-body. Ventrals 240–296. Anal scale divided. Subcaudals 15–25, all divided. **Habitat and range.** Occurs in seasonally dry tropical woodlands and open eucalypt forests from the Ord River drainage, WA, to ne. NT, including Melville and Bathurst Islands. **Behavior.** Nocturnal. A burrowing species observed active on the surface only at night, often after rainstorms. Shelters beneath logs and rocks. Raises parts of its body in loops off the ground if disturbed. **Identification.** *Vermicella multifasciata* is readily identified by its distinctive body shape and coloration, with 77–109 white rings that in some specimens are formed by white spots arranged in a band. It overlaps in distribution with *V. intermedia*, which has 50–53 white rings. **Conservation.** IUCN status: Least Concern. No known major threats.

Northern Bandy Bandy (*Vermicella multifasciata*), Victoria River Downs, NT

Cape York Bandy Bandy
Vermicella parscauda

TL 390 mm. **Lethality.** Potentially dangerous. **Description.** Body small and robust with a short, blunt-tipped tail. Dorsum with alternating black and narrower white rings along entire body and tail; 51–89 white rings. Head short, not distinct from neck, with rounded snout. Ventral area dark or mottled. **Scalation.** Dorsal scales smooth, glossy, and in 15 rows at mid-body. Ventrals 213–230. Anal scale divided. Subcaudals 27, all divided. **Habitat and range.** Occurs in woodlands on heavy red soils and apparently is restricted to an area between Weipa and Mapoon on Cape York Peninsula, QLD. **Behavior.** A poorly known burrowing species, described only in 2018. **Identification.** *Vermicella parscauda* is readily identified by its distinctive body shape and coloration, with 51–89 thin white rings. It overlaps in distribution with *V. annulata*, which has 36–38 white rings. **Conservation.** IUCN status: Data Deficient. May be at risk from large-scale mining operations.

Cape York Bandy Bandy (*Vermicella parscauda*), Oyala Thumotang National Park, Cape York Peninsula, QLD

Pilbara Bandy Bandy
Vermicella snelli

TL 505 mm. **Lethality.** Potentially dangerous. **Description.** Body medium-size and robust with a short, blunt-tipped tail. Dorsum with broad, alternating black and white rings along entire body and tail; 48–64 white rings. Head short, not distinct from neck, with rounded snout. Ventral area with black and white mottling. **Scalation.** Dorsal scales smooth, glossy, and in 15 rows at mid-body. Ventrals 262–302. Anal scale divided. Subcaudals 12–30, all divided. **Habitat and range.** Occurs in shrublands, hummock grasslands, and stony ranges in the Pilbara region of WA. **Behavior.** Nocturnal. A burrowing species observed active on the surface only at night, often after rainstorms. Shelters beneath rocks and in leaf litter. Raises parts of its body in loops off the ground if disturbed. **Identification.** *Vermicella snelli* is readily identified by its distinctive body shape and coloration, with 48–64 white rings, and restricted range in the Pilbara region. **Conservation.** IUCN status: Least Concern. No known major threats.

Pilbara Bandy Bandy (*Vermicella snelli*), Hamersley Ranges, WA

Centralian Bandy Bandy
Vermicella vermiformis

TL 525 mm. **Lethality.** Potentially dangerous. **Description.** Body medium-size and robust with a short, blunt-tipped tail. Dorsum with broad, alternating black and white rings along entire body and tail; 41–45 white rings. Head short, not distinct from neck, with rounded snout. Ventral area with black and white rings. **Scalation.** Dorsal scales smooth, glossy, and in 15 rows at mid-body. Ventrals 263–281. Anal scale divided. Subcaudals 12–30, all divided. **Habitat and range.** Occurs in stony ranges and open eucalypt forests to tropical woodlands; records are from around Alice Springs and Roper River in the NT and the s. Gulf of Carpentaria in QLD. **Behavior.** Nocturnal. A burrowing species observed active on the surface only at night, often after rainstorms. Shelters beneath logs and rocks. Raises parts of its body in loops off the ground if disturbed. **Identification.** *Vermicella vermiformis* is readily identified by its distinctive body shape and coloration, with 41–45 white rings. It may overlap in distribution with *V. annulata*, which has 36–38 white rings. **Conservation.** IUCN status: Least Concern. No known major threats.

Centralian Bandy Bandy (*Vermicella vermiformis*), Alice Springs, NT

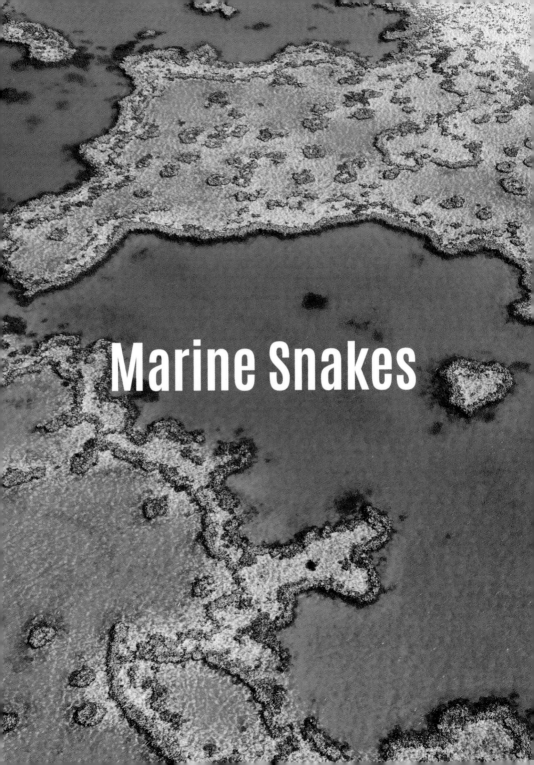

Marine Snakes

Dangerously Venomous Sea Snakes

Twenty-two species (one represented by a subspecies) of viviparous Australian sea snakes are capable of delivering a bite to a human that, if untreated, has a high likelihood of a fatal outcome.

Throughout the world, there are about seventy species of sea snakes (viviparous marine elapids) found in tropical and subtropical waters of the Indian Ocean and the Pacific Ocean, from the e. coast of Africa to the Gulf of Panama. Most species occur in the Indo-Malayan archipelago, the China Sea, Indonesia, and the Australasian region. The viviparous sea snakes originated in Australia, having descended from the country's endemic front-fanged terrestrial snakes.

The group has since radiated in shallow-water marine habitats throughout the Indo-Pacific, where sixty-two species in seven genera are now recognized. Australia supports the world's highest recorded diversity and endemism; more than 35% of the described viviparous sea snake species have been recorded in its waters, and five species are nationally endemic: *Aipysurus apraefrontalis* (Short-nosed Sea Snake), *A. foliosquama* (Leaf-scaled Sea Snake), *A. fuscus* (Dusky Sea Snake), *Ephalophis greyae* (Mangrove Sea Snake), and *Hydrophis donaldi* (Donald's Sea Snake). Over the past 50 years, eight new species have been described from or adjacent to Australian waters, but large areas remain much understudied. Twenty-nine species and one subspecies occur in Australian waters, of which twenty-two are rated as dangerously venomous.

The following eight species occur in Australian waters and generally are considered not dangerous, mostly because their venom apparatus is greatly reduced. Some feed primarily on fish eggs, others on fishes, with some targeting gobies.

Short-nosed Sea Snake	*Aipysurus apraefrontalis*
Leaf-scaled Sea Snake	*Aipysurus foliosquama*
Mosaic Sea Snake	*Aipysurus mosaicus*
Common Turtle-headed Sea Snake	*Emydocephalus annulatus*
West Coast Turtle-headed Sea Snake	*Emydocephalus oriarus*
Mangrove Sea Snake	*Ephalophis greyae*
Black-ringed Mangrove Snake	*Hydrelaps darwiniensis*
Northern Mangrove Sea Snake	*Parahydrophis mertoni*

Genus *Aipysurus*
Sea Snakes

The eight species of medium-size to large snakes of this genus occur mostly in tropical waters of n. Australia. They are considered to inhabit shallow seas, where they forage on the bottom for fishes. They have laterally expanded ventral scales, each at least three times as large as the adjacent lateral body scales and usually with a distinct median keel. Some species are not seen often, while others aggregate in large numbers on particular reefs. They are live-bearers.

Three species are considered not dangerous, and five are dangerously venomous. The venom contains neurotoxins; sea snake antivenom is used to neutralize bites from these species.

Dubois' Sea Snake
Aipysurus duboisii

TL 1.14 m. **Lethality.** Dangerously venomous. **Description.** Body large, robust, and moderately slender with a laterally compressed, paddle-shaped tail. Cream to white with gray or black bands, and scales tipped in white forming a reticulated pattern. Head small, with a pointed snout. Ventral area cream to white. **Scalation.** Dorsal scales smooth, imbricate, weakly keeled, and in 19 rows at mid-body. Ventrals 154–181. Anal scale divided. Subcaudals 25–30, all single. **Habitat and range.** Occurs in coral reefs, mudflats, and seagrass beds from n. NSW to nw. WA. **Behavior.** Nocturnal and crepuscular. Forages in the evening for fishes. Litters of 2–7 young recorded. **Identification.** *Aipysurus duboisii* has all fragmented (asymmetrical) head scalation and 19 mid-body scale rows. **Conservation.** IUCN status: Least Concern. Injury or mortality in trawl-fishing nets is a threat.

Dubois' Sea Snake (*Aipysurus duboisii*), Exmouth, WA

Dubois' Sea Snake (*Aipysurus duboisii*), head, Exmouth, WA

Dusky Sea Snake
Aipysurus fuscus

TL 940 mm. **Lethality.** Dangerously venomous. **Description.** Body medium-size, robust, and moderately short with a laterally compressed, paddle-shaped tail. Dark chocolate brown to blackish brown and weakly patterned. Head small, with a slightly pointed snout. Ventral area dark brown. **Scalation.** Dorsal scales smooth, imbricate, weakly keeled, and in 17–19 rows at mid-body. Ventrals 156–172. Anal scale divided. Subcaudals 24–37, all single. **Habitat and range.** Occurs in coral reefs and reef flats of Ashmore and Scott Reefs in nw. WA. **Behavior.** Predominantly nocturnal. Feeds on gobies and other fishes. **Identification.** *Aipysurus fuscus* has only some fragmented head scalation and 17–19 mid-body scale rows, and it is restricted to Ashmore and Scott Reefs. **Conservation.** IUCN status: Endangered. Threats are largely unknown; however, declines are possibly due to habitat degradation from coral bleaching and decline of ecosystem health.

Dusky Sea Snake (*Aipysurus fuscus*), Ashmore Reef, WA

Olive Sea Snake
Aipysurus laevis

TL 1.7 m. **Lethality.** Dangerously venomous. **Description.** Body large and robust with a laterally compressed, paddle-shaped tail. Purplish brown to gray-brown or white with a golden or dark brown head and lighter-spotted scales on head and body. Head small, with a pointed snout. Ventral area gray-brown to white. **Scalation.** Dorsal scales smooth, imbricate, and in 21–25 rows at mid-body. Ventrals 142–156. Anal scale divided. Subcaudals 22–30, all single. **Habitat and range.** Widespread through coral reefs and rocky areas from n. NSW to Exmouth region of WA. Recorded in high densities in some reefs. **Behavior.** Diurnal and nocturnal. Feeds on a wide variety of fishes and occasionally fish eggs. Litters of 1–5 young recorded. **Identification.** *Aipysurus laevis* has mostly large and symmetrical, fragmented head scalation and smooth imbricate scales in 21–25 mid-body scale rows. **Conservation.** IUCN status: Least Concern. This species is strongly associated with coral reefs, and the degradation of this habitat is likely to pose a threat to its persistence.

Olive Sea Snake (*Aipysurus laevis*), Broome, WA

Olive Sea Snake (*Aipysurus laevis*), swimming, Yongala, QLD

Shark Bay Sea Snake
Aipysurus pooleorum

TL 1.14 m. **Lethality.** Dangerously venomous. **Description.** Body large and robust with a laterally compressed, paddle-shaped tail. Dark brown or purplish brown with pale lateral bars. Head small, with a pointed snout. Ventral area purplish brown. **Scalation.** Dorsal scales smooth, imbricate, and in 20–23 rows at mid-body. Ventrals 146–159. Anal scale divided. Subcaudals 25–33, all single. **Habitat and range.** Inhabits coral reefs and seagrass beds in the Shark Bay area of WA. **Behavior.** Nocturnal and diurnal. Feeds on small fishes. **Identification.** *Aipysurus pooleorum* has mostly large and symmetrical head scalation and 20–23 mid-body scale rows, and is restricted to the Shark Bay area. **Conservation.** IUCN status: Data Deficient. Threats include injury and mortality in trawl-fishing nets.

Shark Bay Sea Snake (*Aipysurus pooleorum*), Shark Bay, WA

Shark Bay Sea Snake (*Aipysurus pooleorum*), head, Shark Bay, WA

Brown-lined Sea Snake
Aipysurus tenuis

TL 1.3 m. **Lethality.** Dangerously venomous. **Description.** Body large and robust with a laterally compressed, paddle-shaped tail. Cream to gray or brown with scales tipped in dark brown forming longitudinal lines and specks. Head darker than body, small, with a pointed snout. Ventral area cream to pale gray. **Scalation.** Dorsal scales smooth, imbricate, and in 19 rows at mid-body. Ventrals 185–194. Anal scale divided. Subcaudals 35–40, all single. **Habitat and range.** Inhabits coral and rocky reefs in nw. WA from Broome to the Dampier Peninsula area. **Behavior.** Recorded eating small fishes. Presumably live-bearing. **Identification.** *Aipysurus tenuis* has fragmented head scalation and 19 mid-body scale rows, and is recorded only from Broome to the Dampier Peninsula. **Conservation.** IUCN status: Data Deficient. Threats include injury and mortality in trawl-fishing nets.

Brown-lined Sea Snake (*Aipysurus tenuis*), Broome, WA

Genus *Hydrophis*
Sea Snakes

This is the largest genus of sea snakes, represented in Australia by seventeen species (one of which is represented by a subspecies) of medium-size to large snakes occurring mostly in tropical waters of n. Australia. Members of this genus vary considerably in form and structure and are characterized by small ventral scales less than twice the width of adjacent body scales. They occur in both deep and shallow waters. Some species are associated with coral reefs while others live in river mouths, estuaries, and mangroves. They are live-bearers.

All Australian species of *Hydrophis* are considered dangerously venomous. The venom contains neurotoxins; sea snake antivenom is used to neutralize bites from these species.

Black-headed Sea Snake
Hydrophis atriceps

TL 1 m. **Lethality.** Dangerously venomous. **Description.** Body medium-size and robust, with a slender forebody and a laterally compressed, paddle-shaped tail. Yellowish brown with a dark head and prominent darker bands and dorsal blotches. Head small, with moderately pointed snout. Ventral area yellowish brown. **Scalation.** Dorsal scales imbricate and in 35–49 rows at mid-body. Ventrals 371–392. Anal scale divided. Subcaudals 47–59, all single. **Habitat and range.** Inhabits coral and rocky reefs in waters around Darwin in the NT. **Behavior.** Nocturnal. Recorded eating small eels and hole-dwelling fishes. Litters of 1–7 young recorded. **Identification.** *Hydrophis atriceps* has a dark head, 35–49 mid-body scale rows, and a very limited range. **Conservation.** IUCN status: Least Concern. Rarely caught in trawl-fishing nets.

Black-headed Sea Snake (*Hydrophis atriceps*), South China Sea, Vietnam

Belcher's Sea Snake
Hydrophis belcheri

TL 1 m. **Lethality.** Dangerously venomous. **Description.** Body medium-size and robust, with a slender forebody and a laterally compressed, paddle-shaped tail. Grayish with paler cream and yellowish narrow bands. Head small and indistinct, with a moderately pointed snout. Ventral area yellowish to pale gray. **Scalation.** Dorsal scales imbricate and in 32–36 rows at mid-body. Ventrals 278–313. Anal scale divided. Subcaudals 28–43, all single. **Habitat and range.** Inhabits coral reefs in waters of the Arafura Sea, between the NT, Australia, and w. New Guinea. **Behavior.** Nocturnal. Habits are poorly known. Recorded eating small eels. Litters of 2–4 young recorded. **Identification.** *Hydrophis belcheri* can be distinguished from other sea snakes by its coloration, 32–36 mid-body scale rows, and limited range. **Conservation.** IUCN status: Data Deficient. Under direct threat from being taken as bycatch in trawl-fishing nets.

Belcher's Sea Snake (*Hydrophis belcheri*), holotype, New Guinea

Dwarf Sea Snake
Hydrophis caerulescens

TL 800 mm. **Lethality.** Dangerously venomous. **Description.** Body medium-size and relatively robust, with a slender forebody and a laterally compressed, paddle-shaped tail. Blue-gray with a blackish head and darker bands and dorsal blotches. Head small, with a moderately pointed snout. Ventral area blue-gray. **Scalation.** Dorsal scales imbricate and in 38–54 rows at mid-body. Ventrals 253–334. Anal scale divided. Subcaudals 40–52, all single. **Habitat and range.** Inhabits river mouths, mangroves, and estuaries in waters of the se. Gulf of Carpentaria, n. Australia. Also, a single specimen recorded from the Fitzroy River in e. QLD. **Behavior.** Nocturnal. Recorded eating small eels and hole-dwelling fishes. Litters of 2–15 young recorded.

Identification. *Hydrophis caerulescens* can be distinguished from other sea snakes by its coloration, 38–54 mid-body scale rows, and limited range. **Conservation.** IUCN status: Least Concern. This species is rarely taken as bycatch in trawl-fishing nets.

Dwarf Sea Snake (*Hydrophis caerulescens*), Mumbai, India

Cogger's Sea Snake
Hydrophis coggeri

TL 1.2 m. **Lethality.** Dangerously venomous. **Description.** Body large and robust, with a slender forebody and a laterally compressed, paddle-shaped tail. Cream to brown with prominent dark gray to black bands. Head small, with a moderately pointed snout. Ventral area cream to brown. **Scalation.** Dorsal scales imbricate and in 29–34 rows at mid-body. Ventrals 280–360. Anal scale divided. Subcaudals 37, all single. **Habitat and range.** Inhabits deeper waters around edges of the Ashmore and Scott Reefs off nw. WA. **Behavior.** Nocturnal. Recorded eating snake eels. Litters of 3–8 young recorded. **Identification.** *Hydrophis coggeri* has distinctive coloration, 29–34 mid-body scale rows, and a limited range. **Conservation.** IUCN status: Least Concern. Under threat from being taken as bycatch in trawl-fishing nets.

Cogger's Sea Snake (*Hydrophis coggeri*), Fannie Bay, NT

Spine-bellied Sea Snake
Hydrophis curtus

TL 1 m. **Lethality.** Dangerously venomous. **Description.** Body medium-size and robust with a laterally compressed, paddle-shaped tail. Olive brown to gray, sometimes with a series of darker bands, which are more conspicuous in juveniles. Head large and distinct from neck, with moderately pointed snout. Ventral area cream to pale yellow. **Scalation.** Dorsal scales juxtaposed, squarish or hexagonal, and in 23–45 rows at mid-body. In adults, scales on the lower flanks much larger, with tubercles or spines, most prominent in males. Ventrals 110–240. Anal scale divided. Subcaudals 239–243, all single. **Habitat and range.** Inhabits reefs and estuaries in tropical waters of nw. WA (including Ashmore Reef), the NT, and QLD. **Behavior.** Mainly nocturnal. Recorded eating a variety of fishes. Litters of up to 15 young recorded. **Identification.** *Hydrophis curtus* has squarish or hexagonal dorsal scales in 23–45 mid-body rows. The lower lateral scales are larger than the dorsal scales and contain enlarged tubercles, keels, or spines, most prominent in adult males. **Conservation.** IUCN status: Least Concern. Under threat from being taken as bycatch by fisheries.

Spine-bellied Sea Snake (*Hydrophis curtis*), Middle Beach, NT

Spine-bellied Sea Snake (*Hydrophis curtis*), Potuvil, Sri Lanka

Fine-spined Sea Snake
Hydrophis czeblukovi

TL 1.25 m. **Lethality.** Dangerously venomous. **Description.** Body large and robust, with a slender neck and strongly, laterally compressed hind body and paddle-shaped tail. Yellowish brown with lighter-colored prominent hexagonal markings over the dorsal and lateral areas of the body. Head small, darker than body, with a moderately pointed snout. Ventral area gray to black. **Scalation.** Dorsal scales juxtaposed, each with a short keel, and in 51–58 rows at mid-body. Ventrals 288–324. Anal scale divided. Subcaudals 45–55, all single. **Habitat and range.** Recorded from Shark Bay, WA, and deeper waters off the continental shelf in nw. Australia and the Arafura Sea in n. Australia. **Behavior.** Nocturnal. A poorly known species, recorded eating small eels. **Identification.** *Hydrophis czeblukovi* has prominent geometrical markings and 51–58 mid-body scale rows. **Conservation.** IUCN status: Data Deficient. Rarely caught in trawl-fishing nets.

Fine-spined Sea Snake (*Hydrophis czeblukovi*), Exmouth, WA

Donald's Sea Snake
Hydrophis donaldi

TL 900 mm. **Lethality.** Dangerously venomous. **Description.** Body medium-size and relatively slender with a laterally compressed, paddle-shaped tail. Yellowish brown with darker brownish bands along the entire body, darker posteriorly. Head small, with a moderately pointed snout. Ventral area yellowish brown. **Scalation.** Dorsal scales imbricate, strongly spinose, and in 33–35 rows at mid-body. Ventrals 246–288. Anal scale divided. Subcaudals 42–51, all single. **Habitat and range.** Known only from shallow estuaries over seagrass and mudflats at the mouths of the Mission River and Hey Creek, where they connect to Albatross Bay in the Weipa coastal area of QLD. **Behavior.** Habits are poorly known; recorded as nocturnal and considered to be live-bearing. **Identification.** *Hydrophis donaldi* has strongly spinose scalation, 33–35 mid-body scale rows, and a very limited distribution. **Conservation.** IUCN status: Data Deficient. No known major threats to this species.

Donald's Sea Snake (*Hydrophis donaldi*), Weipa, QLD

Elegant Sea Snake
Hydrophis elegans

TL 2 m. **Lethality.** Dangerously venomous. **Description.** Body large and elongate, with a slender neck and forebody, robust hind body, and laterally compressed, paddle-shaped tail. Coloration variable, from pale gray to brown, with darker bands. Juveniles have much brighter pattern than adults, with defined black markings. Head small, with a moderately pointed snout. Ventral area cream to white. **Scalation.** Dorsal scales smooth, imbricate, and in 37–49 rows at mid-body. Ventrals 345–432. Anal scale divided. Subcaudals 36–43, all single. **Habitat and range.** Commonly encountered in coral and rocky reefs in waters of e., n., and w. Australia from north of Sydney, NSW, around the continent to the sw. coast of WA. **Behavior.** Nocturnal. Recorded eating small eels and hole-dwelling fishes. Litters of up to 30 young recorded. **Identification.** *Hydrophis elegans* has 35–55 darker crossbands along the body and 37–49 mid-body scale rows. **Conservation.** IUCN status: Least Concern. Under threat from being taken as bycatch in trawl-fishing nets.

Elegant Sea Snake (*Hydrophis elegans*), Ningaloo, WA

Spectacled Sea Snake
Hydrophis kingii

TL 1.9 m. **Lethality.** Dangerously venomous. **Description.** Body large and robust, with a slender forebody and a laterally compressed, paddle-shaped tail. Gray to cream, with numerous dark bands extending halfway down each side; first dark band separated from head by thin whitish bar. Head small and black, usually with a white ring around each eye, with moderately pointed snout. Ventral area black. **Scalation.** Dorsal scales imbricate, keeled, and in 36–40 rows at mid-body. Ventrals 299–360. Anal scale divided. Subcaudals 36–43, all single. **Habitat and range.** Inhabits deeper waters from north of Coffs Harbor, NSW, across n. Australia to Barrow Island, WA. **Behavior.** Nocturnal. Probably feeds on small eels and hole-dwelling fishes. Litters of 1–8 young recorded. **Identification.** *Hydrophis kingii* has a small black head, a white ring around each eye, and 36–40 mid-body scale rows. **Conservation.** IUCN status: Least Concern. Under threat from injury and mortality in trawl-fishing nets.

Spectacled Sea Snake (*Hydrophis kingii*), Eighty Mile Beach, WA

Macdowell's Sea Snake
Hydrophis macdowelli

TL 1 m. **Lethality.** Dangerously venomous. **Description.** Body medium-size and robust, with a very slender forebody and a laterally compressed, paddle-shaped tail. Cream with dark blotches or bands along the body becoming complete rings anteriorly. Head extremely small and dark olive or black, with a moderately pointed snout. Ventral area pale yellow. **Scalation.** Dorsal scales weakly imbricate and in 35–42 rows at mid-body. Ventrals 235–290. Anal scale divided. Subcaudals 36–44, all single. **Habitat and range.** Recorded from coral reefs and sandy estuaries from se. QLD north around the continent to the Pilbara area of WA. **Behavior.** Mostly nocturnal. Recorded eating small eels and other long-bodied fishes. Litters of 2 or 3 young recorded. **Identification.** *Hydrophis macdowelli* has an extremely slender forebody, a very small, dark head, a thickset hind body, and 35–42 mid-body scale rows. **Conservation.** IUCN status: Least Concern. Under threat from injury and mortality in trawl-fishing nets.

Macdowell's Sea Snake (*Hydrophis macdowelli*), Shark Bay, WA

Olive-headed Sea Snake
Hydrophis major

TL 1.6 m. **Lethality.** Dangerously venomous. **Description.** Body large, robust, and consistently thickset, with a laterally compressed, paddle-shaped tail. Pale gray to olive with broad, dark crossbands. Juveniles more boldly marked than adults, with distinct black markings. Head short and stout, with a moderately rounded snout. Ventral area pale gray with scales tipped dark gray. **Scalation.** Dorsal scales imbricate, strongly keeled on the neck, and in 37–45 rows at mid-body. Ventrals 216–273. Anal scale divided. Subcaudals 39–43, all single. **Habitat and range.** Inhabits coral and rocky reefs, estuaries, and seagrass beds from north of Sydney around n. Australia to Bunbury, on the sw. coast of WA. **Behavior.** Nocturnal. Recorded feeding on small fishes. Litters of 6–12 young recorded. **Identification.** *Hydrophis major* has a dorsal pattern with 24–35 dark crossbands and 37–45 mid-body scale rows. It is similar to *H. stokesii*, but that species has longitudinally divided ventral scales. **Conservation.** IUCN status: Least Concern. Under threat from injury and mortality in trawl-fishing nets.

Olive-headed Sea Snake (*Hydrophis major*), juvenile, Broome, WA

Olive-headed Sea Snake (*Hydrophis major*), K'gari (Fraser Island), QLD

Spotted Sea Snake
Hydrophis ocellatus

TL 1.5 m. **Lethality.** Dangerously venomous. **Description.** Body large, robust, and heavily built, with a laterally compressed, paddle-shaped tail. Blue-gray with darker crossbands or blotches along the dorsum and a series of pale ocelli with darker borders along the sides. Head relatively small, with a moderately pointed snout. Ventral area pale cream to white. **Scalation.** Dorsal scales imbricate and in 39–59 rows at mid-body. Ventrals 236–336. Anal scale divided. Subcaudals 38–52, all single. **Habitat and range.** Inhabits coral and rocky reefs as well as turbid inshore waters and estuaries. Recorded from TAS up the e. coast and around n. Australia to sw. WA. **Behavior.** Nocturnal. Feeds on a variety of fishes. Litters of 1–6 young recorded. **Identification.** *Hydrophis ocellatus* has 30–60 broad dark bands or dorsal blotches, a series of large, pale ocelli marking the flanks, and 39–59 mid-body scale rows. **Conservation.** IUCN status: Least Concern. Under threat from being taken as bycatch in trawl-fishing nets.

Spotted Sea Snake (*Hydrophis ocellatus*), Broome, WA

Pacific Sea Snake
Hydrophis pacificus

TL 1.4 m. **Lethality.** Dangerously venomous. **Description.** Body large and robust, with a slender forebody and a laterally compressed, paddle-shaped tail. Dark gray above, sharply delineated at mid-flank from paler ventral color. Prominent darker bands along the body. Juveniles with clear black markings. Head large, with a moderately pointed snout. Ventral area yellowish. **Scalation.** Dorsal scales imbricate and in 39–49 rows at mid-body. Ventrals 320–430. Anal scale divided. **Habitat and range.** Inhabits coral and rocky reefs in waters of the e. Arafura Sea and Gulf of Carpentaria off n. Australia. **Behavior.** Nocturnal. Recorded feeding on small fishes. Litters of up to 17 young recorded. **Identification.** *Hydrophis pacificus* has 49–72 dark bands, 39–49 mid-body scale rows, and a restricted distribution. **Conservation.** IUCN status: Near Threatened. Under threat from being taken as bycatch in trawl-fishing nets.

Pacific Sea Snake (*Hydrophis pacificus*), Gulf of Carpentaria

Horned Sea Snake
Hydrophis peronii

TL 1.23 m. **Lethality.** Dangerously venomous. **Description.** Body large and robust, with a slender forebody and a laterally compressed, paddle-shaped tail. Cream to pale brown or gray, with or without dark crossbands. Head small, with scales strongly keeled, raised scales above eyes, and moderately pointed snout. Ventral area cream to pale brown. **Scalation.** Dorsal scales, each with a central, dark keel, in 21–31 rows at mid-body; scales imbricate on forebody, juxtaposed on hind body. Ventrals 142–203. Anal scale divided. Subcaudals 44, all single. **Habitat and range.** Inhabits coral and rocky reefs, mudflats, and seagrass beds in waters of tropical n. Australia. **Behavior.** Mainly nocturnal. Recorded feeding on gobies and other small fishes. Litters of 1–10 young recorded. Possesses one of the most toxic sea snake venoms known. **Identification.** *Hydrophis peronii* has hornlike spinose scales on head and noticeably (often dark-centered) keeled dorsal scales that are imbricate on the forebody, juxtaposed on the hind body, and in 21–31 mid-body scale rows. **Conservation.** IUCN status: Least Concern. Possibly at risk of being taken as bycatch in trawl-fishing nets.

Horned Sea Snake (*Hydrophis peronii*), Weipa, QLD

Horned Sea Snake (*Hydrophis peronii*), head

Yellow-bellied Sea Snake
Hydrophis platura platura

TL 1 m. **Lethality.** Dangerously venomous. **Description.** Body medium-size and robust with a laterally compressed, paddle-shaped tail. Bright yellow with distinctive black vertebral stripe and yellow tail with black spots or bars. Head large, with moderately pointed snout. Ventral area yellow. **Scalation.** Dorsal scales juxtaposed and in 49–69 rows at mid-body. Ventrals 264–408. Anal scale divided. Subcaudals 39–51, all single. **Habitat and range.** Inhabits mostly open seas; in Australia recorded from TAS up the e. coast and north around the continent to sw. WA. The world's most widespread snake, it occurs throughout the Indian and Pacific Oceans. **Behavior.** Mainly diurnal. Recorded ambushing small pelagic fishes while sitting among floating debris. Litters of 1–6 young recorded. **Identification.** *Hydrophis platura platura* is a uniquely colored snake and unlikely to be confused with any other sea snake. **Conservation.** IUCN status: Least Concern. No known major threats.

Yellow-bellied Sea Snake (*Hydrophis platura platura*), Noumea, New Caledonia

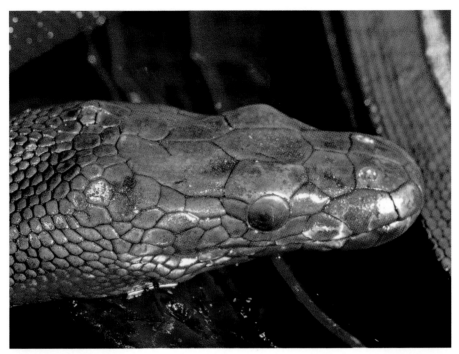

Yellow-bellied Sea Snake (*Hydrophis platura platura*), head

Stokes' Sea Snake
Hydrophis stokesii

TL 2 m. **Lethality.** Dangerously venomous. **Description.** Body large and extremely robust with a laterally compressed, paddle-shaped tail. Yellowish brown to dark brown with darker dorsal bands and blotches. Head large, with a moderately pointed snout and thick neck. Ventral area yellowish brown. **Scalation.** Dorsal scales imbricate and in 54–60 rows at mid-body. Ventrals 252–280, with small ventral scales divided into two strongly overlapping rows. Anal scale divided. Subcaudals 33–36, all single. **Habitat and range.** Inhabits estuaries and rocky reefs in waters from central NSW around n. Australia to nw. WA. **Behavior.** Mainly nocturnal. Recorded eating frogfishes and stonefishes. Litters of 1–14 young recorded. **Identification.** *Hydrophis stokesii* is a large, thickset species, with small ventral scales divided into two strongly overlapping rows, and 54–60 mid-body dorsal scale rows. **Conservation.** IUCN status: Least Concern. Under threat from injury and mortality in trawl-fishing nets.

Stokes' Sea Snake (*Hydrophis stokesii*), Darwin Harbor, NT

Stokes' Sea Snake (*Hydrophis stokesii*), juvenile

Stokes' Sea Snake (*Hydrophis stokesii*), eating gobi fish

Australian Beaked Sea Snake
Hydrophis zweifeli

TL 1.2 m. **Lethality.** Dangerously venomous. **Description.** Body large, strongly compressed laterally, with a laterally compressed, paddle-shaped tail. Cream with gray head and prominent gray bands that fade with age. Head small, with moderately pointed snout and extremely long, dagger-shaped mental scale located on front edge of lower jaw. Ventral area yellowish brown. **Scalation.** Dorsal scales imbricate, each with a short keel, and in 48–56 rows at mid-body. Ventrals 261–313. Anal scale divided. Subcaudals 52, all single. **Habitat and range.** Inhabits estuaries and shallow bays from north of Brisbane, QLD, around n. Australia to near Darwin, NT. **Behavior.** Mainly nocturnal. Recorded eating catfishes, puffer fishes, and prawns. Litters of 1–19 young recorded. **Identification.** *Hydrophis zweifeli* has an extremely long, dagger-shaped mental scale and 48–56 mid-body scale rows. **Conservation.** IUCN status: Data Deficient. Under threat from injury and mortality in trawl-fishing nets.

Australian Beaked Sea Snake (*Hydrophis zwelfeli*), Bali, Indonesia

Dangerously Venomous Sea Kraits

Two species of sea kraits are capable of delivering a bite to a human that, if untreated, has a high likelihood of a fatal outcome.

Genus *Laticauda*
Sea Kraits

A genus of venomous elapid sea snakes with eight species worldwide, *Laticauda* is represented in Australia by two species that occasionally enter the continent's northern waters. Sea kraits are aquatic and amphibious, with wide ventral scales typical of terrestrial snakes for moving on land and flattened paddle-shaped tails for swimming. The nostrils are positioned laterally. Unlike totally aquatic live-bearing sea snakes, sea kraits are oviparous and venture on to land to digest prey and lay eggs. Movement to and from land generally coincides with overnight high tides.

Both *Laticauda* species recorded from Australian waters are capable of delivering a bite to a human that, if untreated, has a high likelihood of a fatal outcome. Normally docile, sea kraits are reluctant to bite. Venom contains neurotoxins, and sea snake antivenom is used to neutralize bites from these species.

Yellow-lipped Sea Krait
Laticauda colubrina

TL 1.4 m. **Lethality.** Dangerously venomous. **Description.** Body large, moderately thick, and essentially cylindrical, with a laterally compressed, paddle-shaped tail. Usually blue to blue-gray with 20–65 black rings. Head small and black on top, with snout and upper lip yellow. Ventral area cream to yellow. **Scalation.** Dorsal scales smooth, imbricate, and in 21–25 rows at mid-body. Ventrals 210–250. Anal scale divided. Subcaudals 25–50, all divided. **Habitat and range.** Aquatic and amphibious, associated with coral reefs but also found in mangrove swamps and seagrass beds. Occurs from the Andaman Islands of India to the w. Pacific; very occasionally recorded in Australia. **Behavior.** Mainly nocturnal. Feeds on fishes including crevice-inhabiting eels. Clutches of 5–19 eggs recorded, which are deposited on land in caves or crevices. **Identification.** *Laticauda colubrina* can be distinguished from *L. laticaudata* by the yellow upper lip and 21–25 mid-body scale rows. **Conservation.** IUCN status: Least Concern. Under threat from injury and mortality in trawl-fishing nets and possibly at risk from coastal development and habitat destruction.

Yellow-lipped Sea Krait (*Laticauda colubrina*), Tavarua Islands, Fiji

Blue-banded Sea Krait
Laticauda laticaudata

TL 1 m. **Lethality.** Dangerously venomous. **Description.** Body medium-size, moderately thick, and essentially cylindrical, with a laterally compressed, paddle-shaped tail. Usually blue to blue-gray with 25–70 black bands or rings. Head small and black on top, with snout and upper lip brown. Ventral area cream to yellow. **Scalation.** Dorsal scales smooth, imbricate, and in 19 rows at mid-body. Ventrals 225–245. Anal scale divided. Subcaudals 25–50, all divided. **Habitat and range.** Aquatic and amphibious, associated with coral reefs but also found in mangrove swamps and seagrass beds. Occurs from Bay of Bengal, India, north to Japan and east to islands of w. Pacific; very occasionally recorded in Australia. **Behavior.** Mainly nocturnal. Feeds on fishes including crevice-inhabiting eels. Clutches of 5–11 eggs recorded, which are deposited on land in caves or crevices. **Identification.** *Laticauda laticaudata* can be distinguished from *L. colubrina* by the brown upper lip and 19 mid-body scale rows. **Conservation.** IUCN status: Least Concern. Under threat from injury and mortality in trawl-fishing nets and possibly at risk from coastal development and habitat destruction.

Blue-banded Sea Krait (*Laticauda laticaudata*), Noumea, New Caledonia

References

Allen, G.E., S.K. Wilson, and G.K. Isbister. 2013. *Paroplocephalus* envenoming: A previously unrecognized highly venomous snake in Australia. Medical Journal of Australia 199 (11), 792–794.

Allen, L., and G. Vogel. 2019. Venomous snakes of Australia and Oceania. Terralog, vol. 18. Frankfurt: Chimaira Buchhandelsgesellschaft.

Bennett, R. 1997. Reptiles and frogs of the Australian Capital Territory. Canberra, ACT: National Parks Association of the ACT.

Broad, A.J., S.K. Sutherland, and A.R. Coulter. 1979. The lethality in mice of dangerous Australian and other snake venom. Toxicon 17: 661–664.

Cann, J. 1986. Snake alive! Snake experts and antidote sellers of Australia. Kenthurst, NSW: Kangaroo Press.

Chapple, D.G., et al. 2019. The action plan for Australian lizards and snakes 2017. Melbourne, VIC: CSIRO Publishing.

Clemann, N., T. Stranks, R. Carland, J. Melville, B. Op den, and P. Robertson. 2017. The Death Adder *Acanthophis antarcticus* (Shaw & Nodder, 1802) in Victoria: Historical records and contemporary uncertainty. Memoirs of Museum Victoria 77: 29–40.

Cogger, H.G. 2014. Reptiles and amphibians of Australia. 7th ed. Melbourne, VIC: CSIRO Publishing.

Coventry, A.J., and P. Robertson. 1991. The snakes of Victoria. Melbourne, VIC: Department of Conservation and Environment.

Eipper, S., and T. Eipper. 2019. A naturalist's guide to snakes of Australia. Sydney, NSW: John Beaufoy Publishing / Australian Geographic.

Gow, G.F. 1982. Australia's dangerous snakes. Melbourne, VIC: Angus and Robertson.

Greer, A. 1997. The biology and evolution of Australian snakes. Sydney, NSW: Surrey Beatty and Sons.

Healy, J., and K.D. Winkel, eds. 2013. Venom: Fear, fascination and discovery. Melbourne, VIC: Medical History Museum, University of Melbourne.

Jackson, T.N.W., et al. 2013. Venom Down Under: Dynamic evolution of Australian elapid snake toxins. Toxins 5 (12): 2621–2655.

Kinghorn, J.R. 1956. The snakes of Australia. 2nd ed. Melbourne, VIC: Angus and Robertson.

Mirtschin, P., and R. Davis. 1982. Dangerous snakes of Australia. Melbourne, VIC: Rigby Publishers.

Mirtschin, P., A.R. Rasmussen, and S.A. Weinstein. 2017. Australia's dangerous snakes. Melbourne, VIC: CSIRO Publishing.

O'Shea, M. 1996. A guide to the snakes of Papua New Guinea. Singapore: Beaumont Publishing.

Rasmussen, A.R., K.L. Sanders, M.L. Guinea, and A.P. Amey. 2014. Sea Snakes in Australian waters (Serpentes: subfamilies Hydrophinae and Laticaudinae): A review with an updated identification key. Zootaxa 3869 (4): 351–371

Sutherland, S.K., and J. Tibballs. 2001. Australian animal toxins. 2nd ed. South Melbourne, VIC, and New York: Oxford University Press.

Shine, R. 1981. Australian snakes: A natural history. Sydney, NSW: Reed.

Swan, G., S. Sadlier, and G. Shea. 2017. A field guide to reptiles of New South Wales. 3rd ed. Sydney, NSW: Reed New Holland.

Swan, M. 2020. Frogs and reptiles of the Murray-Darling Basin. Melbourne, VIC: CSIRO Publishing.

Swan, M., and S. Watharow. 2005. Snakes, lizards and frogs of the Victorian mallee. Melbourne, VIC: CSIRO Publishing.

Watharow, S. 2011. Living with snakes and other Reptiles. Melbourne, VIC: CSIRO Publishing.

Wilson, S., and G. Swan. 2021. A complete guide to reptiles of Australia. 6th ed. Sydney, NSW: Reed New Holland.

Winkel, K., P. Mirtschin, and J. Pearn. 2006. Twentieth century toxinology and antivenom development in Australia. Toxicon (48) 738–754.

Worrell, E. 1952. Dangerous snakes of Australia. Melbourne, VIC: Angus and Robertson.

Glossary

anal scale. The scale just in front of and covering the cloaca; may be single or divided.

anaphylaxis. A severe, sometimes life-threatening allergic reaction.

anterior. Toward the head.

anticoagulant. Something inhibiting or preventing blood clotting.

arboreal. Dwelling in trees.

axillary lymphadenopathy. An enlargement of the axillary (armpit) lymph node.

bifurcated. Divided into two branches or forks.

billabong. A pool of water that is left behind when a river changes course.

biodiversity. The variety of plant and animal life in a particular habitat or area.

buccal gland. A gland that most commonly secretes its contents into the mouth.

cardiotoxin. A toxin that affects the heart.

caudal. Relating to the tail.

cloaca. The common chamber into which the reproductive, urinary, and intestinal ducts open.

coagulant. An agent that promotes clotting of the blood.

coagulopathy. Any alteration of hemostasis resulting in either excessive bleeding or clotting, although typically refers to impaired clot formation.

colubrid. A snake of the family Colubridae.

consumption coagulopathy. A condition affecting the blood's ability to clot.

crepuscular. Active mostly in the twilight hours of evening and early morning.

cryptic. Inconspicuous or secretive.

cryptozoic. Shy, hidden away, or living in concealment.

cytolytic. Referring to the bursting of a cell due to excess water within it.

cytotoxin. A substance having a poisonous effect on cells.

diaphoresis. Sweating, especially of an unusual degree.

diurnal. Active during the day.

dorsal. Of the back.

dorsum. The upper (dorsal) surface of an animal.

elapid. A member of the family Elapidae, the front-fanged venomous snakes.

elliptical. In the shape of an ellipsis; often refers to vertical pupils.

endemic. Restricted to or native to a certain place or region.

endemism. The state of being endemic.

envenomation. The act or result of being injected with venom.

fragmented. Splitting of symmetrically arranged head scales into smaller, irregular scales.

frontal scale. Large scale on the top of the head.

genus. In biology, a principal taxonomic category that ranks above species and below family (plural, genera).

hemolysis. Disruption of the red blood cell membrane, allowing the escape of hemoglobin (adjective, hemolytic).

hemostasis. Normal physiological regulation of bleeding and blood clotting.

hemotoxin. A substance poisonous to red blood cells.

holotype. The actual individual of a species that is used to formally describe it and classify it.

imbricate. Overlapping; in snakes, refers to scales.

iris. The colored circular area surrounding the pupil.

juxtaposed. In scalation, refers to scales that do not overlap.

keeled scales. Scales with a raised ridge.

labial scales. Scales on the lips.

lateral. Relating to the sides.

laterally compressed. Flattened from side to side.

lethal dosage 50 (LD$_{50}$). The dose of a venom that results in the death of 50% of test subjects.

live-bearing. Viviparous; bringing forth live young.

longitudinal. Running along the length of the body.

loreal scale. Scale between nasal and preocular scales; not present in elapid snakes.

lymphadenopathy. An increase in the size of the lymph nodes in response to a particular disease or envenomation.

mallee. A semiarid habitat of various low-growing eucalypt trees.

maxillary. Relating to the maxillae, the two main bones of the upper jaw, which are separated anteriorly by the premaxillary bone in snakes.

mental scale. The single midline scale on the front edge of the lower jaw in lizards and snakes.

mid-body scale rows. The number of dorsal scales (usually counted obliquely) around the middle of the body (see illustration, p. XX).

monotypic. Having only one representative; most often refers to a genus containing a single species.

mulga. A small Australian acacia tree.

myotoxin. A substance poisonous to muscles (adjective, myotoxic).

nasal scale. The scale on the snout that borders or encloses the nostril.

neurotoxin. A substance having a poisonous effect on the nerves and nervous system.

neurotoxicity. Exposure to natural or manufactured toxic substances (neurotoxins) that alter the normal activity of the nervous system.

nocturnal. Active at night.

nominate. The originally described subspecies of a species.

ocelli. Eyelike markings.

oviparous. Egg-laying.

parietal scales. The posterior-most head scales.

polyvalent antivenom. Antivenom made from a number of different immunizing venoms.

posterior. Toward the rear.

prefrontal scale. One of a pair of large scales in front of the frontal scale on the head.

preocular scale. One or more scales along the front edge of the eye.

procoagulant. Toxin or other chemical that stimulates unregulated blood clotting.

proteroglyph. A member of the group of venomous snakes with fangs located in the front of the mouth.

recruitment. In population dynamics, the process by which new individuals are added to a population, whether by birth and maturation or by immigration.

reticulated. A network pattern.

rostral scale. The anterior-most scale on the snout.

sclerophyll. A vegetation type adapted to dry and often hot conditions, in Australia typically consisting of eucalypt, acacia, and banksia species.

species. In biology, a taxonomic category, ranking below genus, consisting of related organisms that share common characteristics and are capable of interbreeding.

spinifex. A genus (*Triodia*) of inland Australian grasses.

spinose. Resembling or having spines.

subcaudal scales. Scales on the underside of the tail.

subspecies. In biology, a taxonomic rank below species used for populations that are different from the main species population.

supralabials. The row of scales that border the upper lip on each side of the rostral scale; also called "upper labials."

supraocular scale. A scale or scales bordering the orbit above the eye.

sympatric. Refers to animal or plant species that inhabit the same geographic region.

syncope. Temporary loss of consciousness caused by a fall in blood pressure.

taxonomy. A system of classification, such as the biological classification of plants and animals (adjective, taxonomic).

temporal scale. One or more scales on the side of the head.

terrestrial. Living on the ground.

toxicity. Degree of poisonousness.

ventral. Of the lower surface or underside.

ventrals. Ventral scales, located on the underside of the body.

vertebral. Of the spine (or vertebral column).

viviparous. Bearing live young.

Photo Credits

Illustrative Photos:

p. ii: Old relief map of Australia (ilbusca/iStock)
p. viii: Sea snake, Sunrise Beach, Queensland (frankiefotografie/iStock)
p. x: Tree ferns, Black Spur Drive, Healesville, Victoria (TonyFeder/iStock)
p. 18: *Notechis scutatus serventyi* (S. Black)
p. 21: Australian Outback (Ingrid_Hendriksen/iStock)
p. 195: Great Barrier Reef (yanjf/iStock)

Species Photos:

L. Allen, p. 49; **S. Black**, p. 41 (bottom), p. 43 (bottom), p. 47 (insert), p. 48 (top), p. 80 (both photos), p. 89; **J. Breedon**, p. 202; **A. Brice**, p. 5, p. 43 (third from top), p. 76 (bottom); **B. Bush**, p. 116, p. 133; **N. Callanan**, p. 227 (bottom); **J. Campbell**, p. 166; **M. Cermak**, p. 10 (all photos), p. 12; **A. Cleary**, p. 79 (main photo); **H. G. Cogger**, p. 200, p. 204, p. 208, p. 214, p. 224, p. 225, p. 231, p. 232; **S. Eipper**, p. 42 (top), p. 48 (bottom), p. 69, p. 74, p. 99, p. 221; **T. Eipper**, p. 43 (top); **J. Farquhar**, p. 26, p. 113, p. 149, p. 153, p. 154, p. 157, p. 179, p. 181, p. 186, p. 187; **N. Gale**, p. 35, p. 68 (top); **P. Ghadekar**, p. 209; **K. Griffiths**, p. 61; **A. Holmes**, p. 189 (insert); **P. Horner**, p. 191, p. 210, p. 211, p. 226; **M. Jackson**, p. 46 (bottom); **S. Mahony**, p. 24, p. 41 (top), p. 86, p. 124, p. 125, p.128; **C. Margetts**, p. 41 (middle); **B. Maryan**, p. 66 (bottom), p. 134, p. 170, p. 177, p. 216, p. 222; **J. Meney**, p. 52, p. 82, p. 105; **National Library of Australia**, p. 16; **G. Parker**, p. 194; **A. Rasmussen**, p. 207; **A. Samuel**, p. 72; **B. Schembri**, p. 66 (top), p. 70, p. 71 (both photos), p. 112, p. 115, p. 119, p. 140, p. 160, p. 174, p. 182, p. 185, p. 190, p. 193; **S. Scott**, p. 6 (both photos), p. 45, p. 46 (top), p. 53, p. 73, p. 121, p. 141, p. 142, p. 143, p. 155, p. 164, p. 169, p. 175, p. 176, p. 183; **O. Sherlock**, p. 38; **R. Sillett**, p. 62; **R. Somaweera**, p. 198, p. 199, p. 201, p. 205, p. 212, p. 213, p. 215, p. 217, p. 218, p. 219, p. 220, p. 223, p. 227 (top), p. 228; **G. Stephenson**, p. 25 (bottom), p. 28, p. 29, p. 31, p. 32, p. 37, p. 42 (middle and bottom), p. 47 (main photo), p. 54, p. 55, p. 56, p. 57, p. 58 (both photos), p. 65 (both photos), p. 67, p. 68 (bottom), p. 75, p. 76 (top), p. 77, p. 78, p. 79 (insert), p. 91, p. 92, p. 95, p. 96, p. 98, p. 104, p. 106, p. 114, p. 126, p. 135, p. 137, p. 138, p. 139, p. 147, p. 150, p. 151, p. 163, p. 167, p. 180, p. 184, p. 189 (main photo); **P. Street**, p. 118; **S. Subaraj**, p. 203; **J. Sulda**, p. 43 (second from top); **M. G. Swan**, p. 59; **J. Vos**, p. 30 (both photos), p. 33 (both photos), p. 50, p. 101, p. 102, p. 159, p. 172; **G. Wallis**, p. 36; **D. Williams**, p. 27, p. 60; **S. K. Wilson**, p. 25 (top), p. 63, p. 93, p. 97, p. 110, p. 117, p. 144, p. 145; **J. Wright**, p. 129; **A. Zimny**, p. 88, p. 109, p. 120, p. 123, p. 130, p. 131, p. 136, p. 146, p. 162, p. 192.

Species Checklist

Dangerously Venomous Land Snakes

Common Death Adder,	*Acanthophis antarcticus*	24
Barkly Death Adder	*Acanthophis hawkei*	26
Smooth-scaled Death Adder	*Acanthophis laevis*	27
Kimberley Death Adder	*Acanthophis lancasteri*	28
Northern Death Adder	*Acanthophis praelongus*	29
Desert Death Adder	*Acanthophis pyrrhus*	30
Top End Death Adder	*Acanthophis rugosus*	31
Pilbara Death Adder	*Acanthophis wellsi*	32
Pygmy Copperhead	*Austrelaps labialis*	35
Highlands Copperhead	*Austrelaps ramsayi*	36
Lowlands Copperhead	*Austrelaps superbus*	37
Tiger Snake	*Notechis scutatus*	40
Inland Taipan	*Oxyuranus microlepidotus*	45
Coastal Taipan	*Oxyuranus scutellatus scutellatus*	47
Papuan Taipan	*Oxyuranus scutellatus canni*	49
Western Desert Taipan	*Oxyuranus temporalis*	50
Mulga Snake or King Brown Snake	*Pseudechis australis*	52
Spotted Mulga Snake	*Pseudechis butleri*	54
Collett's Snake	*Pseudechis colletti*	55
Spotted Black Snake	*Pseudechis guttatus*	57
Eastern Pygmy Mulga Snake	*Pseudechis pailsei*	59
Papuan Black Snake	*Pseudechis papuanus*	60
Red-bellied Black Snake	*Pseudechis porphyriacus*	61
Western Pygmy Mulga Snake	*Pseudechis weigeli*	63
Dugite	*Pseudonaja affinis*	65
Strap-snouted Brown Snake	*Pseudonaja aspidorhyncha*	67
Speckled Brown Snake	*Pseudonaja guttata*	70
Peninsula Brown Snake	*Pseudonaja inframacula*	72
Ingram's Brown Snake	*Pseudonaja ingrami*	74
Western Brown Snake	*Pseudonaja mengdeni*	75
Ringed Brown Snake	*Pseudonaja modesta*	77
Northern Brown Snake	*Pseudonaja nuchalis*	78
Eastern Brown Snake	*Pseudonaja textilis*	79
Rough-scaled Snake	*Tropidechis carinatus*	82

Medically Significant Venomous Land Snakes

Potentially Dangerous Venomous Land Snakes

Dangerously Venomous Sea Snakes

Dangerously Venomous Sea Kraits

Index

About the Author

Mike Swan has enjoyed a lifelong interest in herpetology and has traveled throughout Australia and internationally in pursuit of his interest. He has worked as a senior herpetofauna keeper with Melbourne Zoo and Healesville Sanctuary in Victoria and been part of numerous captive-breeding programs. He also has extensive experience with Australia's dangerous snakes, an area of special interest. He is a keen photographer and has written numerous articles, papers, and books about reptiles and frogs.

Mike is currently the coordinator of the Lilydale High School, Victoria, reptile collection, the largest school collection of reptiles and frogs in Australia.

Notes

Notes

Notes